BIOLOGICAL EFFECTS OF LOW LEVEL EXPOSURES to CHEMICALS and RADIATION

Editor
Edward J. Calabrese

CRC Press
Taylor & Francis Group
Boca Raton London New York

CRC Press is an imprint of the
Taylor & Francis Group, an **informa** business

First published 1992 by CRC Press
Taylor & Francis Group
6000 Broken Sound Parkway NW, Suite
300 Boca Raton, FL 33487-2742

Reissued 2018 by CRC Press

© 1992 by Taylor & Francis
CRC Press is an imprint of Taylor & Francis Group, an Informa business

No claim to original U.S. Government works

A Library of Congress record exists under LC control number: 91035010

Publisher's Note
The publisher has gone to great lengths to ensure the quality of this reprint but points out that some imperfections in the original copies may be apparent.

Disclaimer
The publisher has made every effort to trace copyright holders and welcomes correspondence from those they have been unable to contact.

ISBN 13: 978-1-138-50672-5 (hbk)
ISBN 13: 978-1-138-55774-1 (pbk)
ISBN 13: 978-1-315-15028-4 (ebk)

Visit the Taylor & Francis Web site at http://www.taylorandfrancis.com and the CRC Press Web site at http://www.crcpress.com

Edward J. Calabrese is a board certified toxicologist who is professor of toxicology at the University of Massachusetts School of Public Health, Amherst. Dr. Calabrese has researched extensively in the area of host factors affecting susceptibility to pollutants, and is the author of more than 270 papers in scholarly journals, as well as 15 books, including *Principles of Animal Extrapolation; Nutrition and Environmental Health*, Vols. I and II; *Ecogenetics; Safe Drinking Water Act: Amendments, Regulations and Standards; Petroleum Contaminated Soils,* Vols. 1, 2, and 3; *Ozone Risk Communication and Management; Hydrocarbon Contaminated Soils*, Volume 1; and *Hydrocarbon Contaminated Soils and Groundwater*, Volume 1. His most recent books include *Multiple Chemical Interactions; Air Toxics and Risk Assessment; Alcohol Interactions with Drugs and Chemicals*; and *Regulating Drinking Water Quality*. He has been a member of the U.S. National Academy of Sciences and NATO Countries Safe Drinking Water committees, and has been appointed to the Board of Scientific Counselors for the Agency for Toxic Substances and Disease Registry (ATSDR). Dr. Calabrese also serves as Chairman of the International Society of Regulatory Toxicology and Pharmacology's Council for Health and Environmental Safety of Soils (CHESS) and Director of the Northeast Regional Environmental Public Health Center at the University of Massachusetts.

Preface

In May 1990 a group of scientists representing several federal agencies, the International Society of Regulatory Toxicology and Pharmacology, the private sector, and academia met to develop a strategy to encourage the study of the biological effects of low level exposures (BELLE) to chemical agents and radioactivity. The meeting was convened because of the recognition that most human exposures to chemical agents and radioactivity are at low levels, yet most toxicological studies assessing potential human health effects deal with exposures to extremely high levels, often orders of magnitude greater than actual exposure. Consequently, risks at low levels are estimated by mathematical modeling, utilizing assumptions about which there is considerable uncertainty.

The organizing committee is committed to the enhanced understanding of low-dose responses of all types, of an expected nature (e.g., linear, sublinear) or of a so-called paradoxical nature. Paradoxical dose-response relationships might include U-shaped dose-response curves, hormesis, and, in some restrictive sense, biphasic dose-response curves. Although there are many scattered reports of such paradoxical responses in the biomedical literature, these responses have not been rigorously assessed, nor have the underlying mechanisms been adequately identified. Laboratory and regulatory scientists have tended to dismiss these paradoxical responses as curiosities or anomalies inconsistent with the conventional paradigm.

The proposed activities of the BELLE committee would focus on dose-response relationships to toxic agents, pharmaceuticals, and natural products studied over wide dosage ranges in in vitro systems and in vivo systems, including human populations. The initial goal of BELLE is the scientific evaluation of the existing literature and of ways to improve research and assessment methods. While the principal emphasis of the committee is to promote an assessment of the scientific foundations of BELLE, the need to assess the regulatory implications of BELLE is recognized and will be incorporated in BELLE activities.

The Advisory Committee (pp. ix–xi) authorized Professor Edward J. Calabrese, School of Public Health, University of Massachusetts, Amherst, to organize a workshop on current knowledge about BELLE, with particular emphasis on the toxicological implications of biological adaptations. This meeting was held on April 30 and May 1, 1991, at the University of Massachusetts. The meeting was designed to help establish a scientific base for future BELLE initiatives. The meeting was attended by the seven invited speakers, the BELLE Advisory Committee, and a number of invited guests from universities, federal agencies, and private sector organizations.

The focus of the invited presentations was on the issue of the toxicological implications of biological adaptations. The selection of topics and speakers was designed to consider critically the concept of hormesis, not only in a broad, conceptual manner (Chapter 1, "Hypotheses on Mammalian Aging, Toxicity, and Longevity Hormesis: Explication by a Generalized Compertz Function," Harold Boxenbaum), but also at molecular and biochemical levels (Chapter 2, "The Role of the 'Stress Protein Response' in Hormesis," Joan Smith-Sonneborn); (Chapter 3, "DNA Repair: As Influenced by Age, Nutrition, and Exposure to Toxic Substances," Ron Hart et al.); (Chapter 4, "Biochemical Mechanisms of Biphasic Dose-Response Relationships: Role of Hormesis," Harihara M. Mehendale). These three mechanistically oriented presentations offered a complementary perspective on the diverse range of molecular mechanisms that can become activated at low levels of toxicant exposure and lead to organismal responses seen in the presentation of Boxenbaum.

Although the major focus of the meeting dealt with responses to low levels of chemical exposures, it should be noted that considerable research has been directed toward the effects of low levels of radiation on biological systems. In fact, in 1987 the journal *Health Physics* published a 1985 conference proceedings on the topic of radiation hormesis. Consequently, Robert Anderson provided a synthesis of current research on this topic in his presentation "Effects of Low-Dose Radiation on the Immune Response" (Chapter 5). Emmanuel Farber (Chapter 6, "Cellular Adaptation as an Important Response During Chemical Carcinogenesis") offered a unifying perspective on how the liver adapts to genetic insults via the formation of hepatocytes that are resistant to subsequent toxic insults due to a diminished capacity to bioactivate xenobiotics and an enhanced capacity to facilitate xenobiotic excretion. The relationship of the resistant hepatocyte theory and modern molecular understandings of the process of carcinogenesis was addressed. It is Farber's opinion that the early and intermediate stages in carcinogenesis, as embodied in the resistant hepatocyte, are principally "physiological-adaptive," and are distinguishable from the progressive stages of carcinogenesis involving frank malignancy.

The final presentation was devoted to biostatistical considerations when designing studies that address issues associated with the biological responses to low doses of chemicals and radiation, as well as issues in the interpretation of the findings from such studies (Chapter 7, "Biostatistical Approaches for Modeling U-Shaped Dose-Response Curves and Study Design Consideration in Assessing the Biological Effects of Low Doses," Tom Downs).

This workshop provided an important benchmark for future BELLE activities. The presentations have indicated that biological systems have an impressive array of adaptations that are turned on in response to various stresses, including physiological stress, as well as exposure to radiation, toxic chemicals, and dietary alterations. Despite the striking findings of some of the presentations, such as that by Hart and colleagues that DNA repair efficiency and fidelity are markedly enhanced in caloric restricted diets, the implications of these findings for human populations remains to be further investigated and established. Nonetheless, this publication of the BELLE program will provide the first of what is hoped to be a series of carefully coordinated and focused reports that will clarify the biological effects of low level exposures to chemical agents and radiation on biological systems and human populations.

BELLE Committee Members
February 1991

Edward J. Calabrese, Ph.D.
Chair, BELLE Advisory Committee
School of Public Health
University of Massachusetts
Amherst, MA 01003

Dr. Mary F. Argus
U.S. EPA – TS-796
401 M Street, SW
Washington, DC 20460

James Robert Beall, Ph.D.
ER–72 GPN
MS–G236
Office of Health and Environmental Research
U.S. Department of Energy
Washington, DC 20545

George C. Becking, Ph.D.
EPCS/IRRV at NIEHS
World Health Organization
P.O. Box 12233
Research Triangle Park, NC 27709

Ralph Cook, M.D.
The Dow Chemical Company
Washington Street
1803 Building
Midland, MI 48674

J. Michael Davis, Ph.D.
Health Scientist
U.S. EPA (MD–52)
Env. Criteria & Assmnt. Office
Research Triangle Park, NC 27711

Max Eisenberg, Ph.D.
Executive Director
Center for Indoor Air Research
1099 Winterson Road, Suite 280
Linthicum, MD 21090

William Farland, Ph.D.
U.S. EPA - RD-689
Office of Health and Environmental Assessment
401 M Street, SW
Washington, DC 20460

James R. Fouts, Ph.D.
Senior Scientific Advisor to
  the Director
NIEHS
P.O. Box 12233
Research Triangle Park, NC 27709

John Frawley, Ph.D.
International Society of Regulatory
  Toxicology and Pharmacology
Hercules, Inc.-Hercules Plaza
1313 Market Plaza
Wilmington, DE 19894

Hank S. Gardner, MSPH
Research Biologist
Health Effects Research Division
U.S. Army Medical Bioengineering
Research and Development Lab
Fort Detrick
Frederick, MD 21701-5010

Ronald W. Hart, Ph.D.
Department of Health & Human Services
Director, National Center for
  Toxicological Research
NCTR Drive
Jefferson, AR 72079

Wallace A. Hayes, Ph.D.
Bowman Gray Technical Center
RJR Nabisco, Inc.
Winston-Salem, NC 27102

Donald H. Hughes, Ph.D.
The Procter & Gamble Company
Research and Development Department
Ivorydale Technical Center
Cincinnati, OH 45217

John Keller, Ph.D.
Life Systems, Inc.
1225 Jefferson Davis Highway
Suite 509
Arlington, VA 22202

Dan Krewski, Ph.D.
Environmental Health Center
Room 109
Tunney's Pasture
Ottawa, Ontario K1AOL2

Leonard Sagan, Ph.D.
Electric Power Research Institute
3412 Hillview Avenue
Palo Alto, CA 94303

Harry Salem, Ph.D.
Toxicology Division, CRDEC
U.S. Army
SMCCR-RST
Aberdeen Proving Grounds, MD 21010-5423

Andrew Sivak, Ph.D.
Health Effects Institute
215 First Street
Cambridge, MA 02142

Richard Dean Thomas, Ph.D.
National Research Council, Tox/Epi
National Academy of Sciences
2101 Constitution Ave.-JH 653
Washington, DC 20418

Contents

BIOLOGICAL EFFECTS OF LOW LEVEL EXPOSURES

to
CHEMICALS and RADIATION

CHAPTER 1

Hypotheses on Mammalian Aging, Toxicity, and Longevity Hormesis: Explication by a Generalized Gompertz Function

Harold Boxenbaum, Marion Merrell Dow, Cincinnati, Ohio

INTRODUCTION

> I work all day, and get half-drunk at night.
> Waking at four to soundless dark, I stare.
> In time the curtain-edges will grow light.
> Till then I see what's really always there:
> Unresting death, a whole day nearer now . . .
>
> Philip Larkin[1]

The triune concepts of aging, toxicity, and longevity hormesis are best integrated through analysis of mortality kinetic data. Consequently, the initial part of this introduction will review basic, relevant concepts in this area.

Assuming an organism has survived to the onset (x) of a time interval (x + Δx), the probability of dying over that interval is termed the *age-specific mortality rate,* and the time associated with that probability is usually taken as the midtime of the interval. When Δx becomes dx, the age-specific mortality rate assumes a theoretical designation, namely, the *hazard function,* h(x). The hazard function is also termed the *instantaneous mortality rate, force of mortality,* and *conditional mortality rate.* It is defined by the following relationship:

$$h(x) = \frac{-d\ [\ln\ S(x)]}{dx} = \frac{-1}{N(x)}\ \frac{dN(x)}{dx} \tag{1}$$

where h(x) = the hazard function at time x
   S(x) = the fraction of individuals surviving to time x
   N(x) = the number of individuals surviving to time x

The Napierian logarithm of the hazard function is termed the *Gompertz function, Gompertz transform* or *transformation,* or simply the *Gompert-*

zian.[2] The latter term was created by George Sacher to honor Benjamin Gompertz (1779–1865) for advancing a rate theory approach to mortality data. The son of an Amsterdam diamond merchant, Gompertz was the first actuarial scientist to report that age-specific mortality increased exponentially in adult human populations.[3,4] Subscribing to what is now known as the "wear-and-tear" theory of aging, Gompertz likened human mortality to the "exhaustions of the receiver of an air pump by strokes repeated at equal intervals of time."

In its most generalized form,[5] the Gompertz function is:

$$G_x = G_o + \phi x \qquad (2)$$

where G_x = the Napierian logarithm of the hazard function (i.e., the Gompertz function/transformation or Gompertzian)

G_o = the vulnerability parameter

ϕx = either a linear or curvilinear function

G_o is an index of the vigor of the genotype in its environment.[5,6] Gauging population inability to withstand endogenous and exogenous mediated injury,[5] it provides the initial condition from which the second law of thermodynamics can impel the population from more ordered to less ordered states.[7,8] The derivative of ϕx specifies the rate at which this progressive instability (and, consequently, mortality) is manifested both from internal and external sources. Thus, G_o and ϕ are each dependent on genotype and environment, as well as their interaction.[9] Not surprisingly, there is considerable variability in both G_o and ϕ across human populations.[10-12] In terms of our discussion, *senescence* and *aging* refer to those processes, in the context of a hospitable, stable environment, that directly impact on G_o and ϕ.

For inbred, eutherian, mammalian laboratory populations housed under good laboratory conditions and kept free of preventable disease, the simplest Gompertz function that frequently characterizes mortality experience is the linear form:[5]

$$G_x = G_o + \alpha x \qquad (3)$$

where α is a first-order mortality rate constant. In fitting Equation 3 — or any Gompertz function — to data, G_x is estimated by taking the Napierian logarithm of the age-specific mortality rate. Age-specific mortality rate, Ω_x, is conveniently calculated by substituting Δ for d in Equation 1:[2,13]

$$\Omega_x = \frac{-\Delta [\ln S(x)]}{\Delta x} \qquad (4)$$

The time at which Ω_x is operative is taken as the midtime of the interval. A variance estimate for [ln Ω_x] may also be readily calculated.[2,14] The reciprocal of this variance estimate is appropriate for weighting the dependent variable in linear or nonlinear least squares analysis (see below).

As a means of investigating the usefulness of Gompertz analysis in the characterization of mammalian aging, toxicity, and longevity hormesis, a series of hypotheses will initially be posited and discussed. This will be followed by a more in-depth discussion of longevity hormesis, including a survey and evaluation of the database. As used here, the term *hypothesis* is applied in a general sense, referring to a proposition or declaration that has not yet attained a high degree of credibility,[15] a conception of what the truth might be.[16] Consequently, a hypothesis is capable of being believed, doubted, or denied.[17] Moreover, some of the hypotheses presented may appropriately be perceived as empirical observations, statements of fact, postulates, paradigms, models, theories, etc. However, these distinctions will not be attempted by the author.

Of paramount interest is the concept of longevity hormesis, characterization of which is facilitated through Gompertz analysis. Unfortunately, even the generic concept of hormesis has trod a tortuous path,[18] with many not believing in its existence. Nonetheless, a historically based definition of hormesis will be posited for heuristic purposes: hormesis is a biological phenomenon in which a beneficial or stimulatory effect is obtained through application of a biologically nonessential agent generally considered detrimental or toxic to the system under scrutiny.

Obviously, this definition is unsettling. For one thing, a stimulatory response may not be beneficial, e.g., induction of microsomal enzymes that increase production of mutagens and carcinogens. Another difficulty arises in reconciling the diverse processes that have been declared hormetic: growth, development, hatching success, length of reproductive life, behavioral parameters, microsomal enzyme induction, fecundity, cancer reduction, disease resistance, viability, respiration, radiation resistance, wound healing, resistance to infection, DNA repair, immune system stimulation, and longevity enhancement.[18] In essence, the concept of hormesis rests upon biological process descriptions, the linchpins of which are subjective judgments as to whether or not the processes are "good" for the organism and/or species.[19] Making matters worse, mechanisms underlying these so-called hormetic responses are either poorly understood or not understood at all.[18,20-22] Despite these problems, there do exist some phenomena, labeled hormetic, that are intriguing because they are counter-intuitive. A good example is growth hormesis in peppermint.[23] Application of low concentrations of the growth retardant phosfon actually stimulates growth of the plant.

Another example is longevity hormesis.[5,18,21,24,25] Longevity hormesis is a phenomenon of unknown mechanism whereby an exogenous agent reversibly acts to reduce age-specific mortality rates (relative to controls), in addition to any other actions elicited by the agent. This is not intended to be a definition of longevity hormesis, but rather a vague description of it. A proper definition awaits delineation of mechanism (see Neafsey[21] for a review of proposed mechanisms). For longevity hormesis to be operative, it is not required that observed (net) age-specific mortality rates be reduced relative to control animals, since toxicity from the agent may predominate and actually bring about an overall increase in age-specific mortality. All that is required is that some hormetic biological activity exists that reduces age-specific mortality rates from values that theoretically would be present in the absence of this activity. In the simplest case, an agent only produces longevity hormesis; that is, toxicity and other effects are completely absent. In this situation, only reductions in age-specific mortality rates would theoretically occur. When this occurs in practice, it is generally accompanied by decreases in age-associated pathologies.[24-26]

Unfortunately, this situation occurs infrequently with data from the literature. More often, both longevity hormetic and toxic actions elicited by an agent are simultaneously observed, mainly because most data come from studies in which toxic effects are being investigated, for example, risk assessment (bioassay) rodent studies. With the exception of antioxidant feeding studies, caloric restriction studies, and physical activity studies, it is difficult to find experiments specifically designed to demonstrate life prolongation in animal populations, especially mammals. And even when these studies are found, one frequently cannot make plots of age-specific mortality rates vs time. Investigators frequently report only mean or median life spans, especially when studying insect populations such as *Drosophila* (fruit flies).

Another confounding variable encountered in literature data is life prolongation caused by reduced caloric intake and/or suppressed weight gain. Administration of exogenous agents, especially toxic ones, and especially those admixed with food, often causes rodents to eat less and gain less weight. This reduced food consumption/suppressed weight gain is associated with an irreversible (permanent) longevity enhancement, unrelated to longevity hormesis.[21,24,25,27,28] The two types of longevity enhancement, that from caloric restriction (irreversible) vs that from longevity hormesis (reversible), must not be confused.

As mentioned earlier, many of the concepts advanced in this chapter will be discussed in the context of hypotheses. Since longevity hormesis is described in terms of the Gompertz function, and since the Gompertz function is dependent upon aging and toxicity, as well as longevity hormesis, it is not possible to omit hypotheses dealing with these biological phenomena. A thorough knowledge of how all these processes affect the Gompertz function is necessary for an understanding of longevity hormesis. Accordingly, the database characterizing the longevity hormesis phenomenon will only be discussed after a proper foundation has been laid.

THE HYPOTHESES

The Moving Finger writes; and, having writ,
Moves on: nor all thy Piety nor Wit
Shall lure it back to cancel half a Line,
Nor all thy Tears wash out a Word of it.

Rubáiyát of Omar Khayyám[29]

Hypothesis I: Mammalian senescence is "progressive instability that arises from a slow and continuing change of constraints as a by-product of energy throughput and action."

This hypothesis, proposed by Yates,[8,30] evinces the view that "entropy accumulation shrinks the viability reserves of the organism and renders it more susceptible to the action of fluctuations." The view that "constraints shall not last," a reiteration of the second law of thermodynamics, subsumes the idea of mammalian senescence and assures that progressive deterioration of bodily faculties and performances will occur "(the moving finger)" as we grow old.[8,16,30] Deterioration occurs because the rate of endogenous damage exceeds the rate of endogenous repair. From the point of view of Asimov, aging is inevitable, because the "significant chemicals of living tissue are rickety and unstable, which is exactly what is needed for life."[31]

Many investigators (see especially Cutler[32]) believe senescence occurs primarily as a result of oxygen metabolism; these reactions produce free radicals, aldehydes, and a wide range of peroxides highly toxic to cells. Cutler[32] speculates that the by-products of oxidations damage vital cellular components, causing cells to drift from their proper state of differentiation to a state of dysdifferentiation. One interesting example of cellular dysdifferentiation is the synthesis of hemoglobin by neuron cells.[32] Bass et al.[33] developed a mortality model along the same lines. They derived a Gompertz-Makeham function by assuming a competition between hypothetical life-prolonging and life-shortening regulatory elements (ultimately won by the latter) interacting by generalized Volterra-type competitive exclusion.

On a systems level, it appears that aging is regulated principally through integrative mechanisms involving mainly the brain, the endocrine glands, and the immune tissues.[34] Collectively, these elements constitute what has come to be known as the neuroendocrinimmune system.

Hypothesis II: Mammalian toxicity is an instability arising from a decay of constraints, instigated by exposure to nonessential, noxious exogenous agents/stimuli or overexposure to essential agents/stimuli.

Armed with Hypotheses I and II, we attempt to differentiate senescence from toxicity. The former arises as the "cost of living" from essential processes; the latter is a consequence of deleterious processes provoked by unnecessary, exogenous insults or exposures. Of course, it is not always easy or possible to distinguish between the two. For example, is anxiety exogenously induced? What dose constitutes overexposure to vitamin C? Are saturated fats more injurious than unsaturated fats, and can consumption of one over the other constitute toxicity? However, regardless of whether instability results from senescence or toxicity, it manifests itself as injury, inferred to be deleterious modification of vital system states.[35]

Hypothesis III: For homogeneous laboratory populations of eutherian mammals housed under uniform, good laboratory conditions and kept free of preventable disease, the hazard function is proportional to the mean level of injury in the population.

This hypothesis comes from Sacher and Trucco;[5,36-39] there exist both a theoretical basis as well as experimental evidence in its support. It is a powerful hypothesis, since it allows the time course of mean population injury to be probed through mortality data. Mathematically, Hypothesis III may be expressed as follows:

$$h(x) = k\ e^{\phi x} \qquad (5)$$

where k = a constant of proportionality
ϕx = the mean level of injury in the population at time x

When multiple injury processes occur simultaneously (e.g., senescence and toxicity), ϕx is a weighted composite. Taking Napierian logarithms of Equation 5, one obtains:

$$\ln\ [h(x)] = \ln k + \phi x = G_x = G_o + \phi x \qquad (6)$$

This is the generalized Gompertz function, previously posited as Equation 2, but now with an interpretation of ϕx.

Application of this hypothesis assumes and requires both genetic and environmental homogeneity, as well as good animal husbandry. Elements of good animal husbandry include proper regulation of temperature, lighting, heating, humidification, air quality, ventilation, hygiene, caging, handling, bedding, water, diet and nutrition, etc.[40] "Preventable disease" refers to communicable disease (e.g., bacterial infection) as well as those caused by unnecessary exposure to exogenous agents.

Hypothesis IV: For homogeneous laboratory populations of eutherian mammals housed under uniform, good laboratory conditions and kept free of preventable disease, the most commonly observed Gompertz function is its linear form.

This hypothesis comes from Sacher.[5] For applicable populations, it specifies that senescent injury accrues at a constant rate. Data consistent with this hypothesis have been observed by Boxenbaum, Neafsey, and colleagues,[14,24,25,28] and the thermodynamic approach of Yates[8] supports this gradualist, uniform view of aging. Although it is possible that another function may be deemed more consistent in the future (e.g., the Weibull function[41,42]), it seems unlikely that it will deviate significantly from the linear Gompertz form.

Hypothesis V: Increases and/or decreases in mammalian injury elicited by nonessential, exogenous agents or stimuli are superimposable with senescent injury.

Once again, we are indebted to Sacher and colleagues[5,36,39,43,44] for the unfolding of this hypothesis. The principle of superimposition may be posited as follows:[45] If y_1 is the system response to an input x_1, and y_2 is the response to input x_2, and k_1 and k_2 are arbitrary coefficients, then the response to input $k_1x_1 + k_2x_2$ is $k_1y_1 + k_2y_2$. Viewed informally, the principle of superimposition requires that the whole be the sum of its individual parts.

With the introduction of this hypothesis, we are able to propose interesting alterations to Gompertz functions that characterize homogeneous populations.[7,14,18,21,24,25,27,28,33,46,47] For purposes of simplification, assume aging or senescent injury accrues in accordance with a linear Gompertz function.

In the first case, assume a single dose of a toxic insult produces an instantaneous and constant residue of irreparable (permanent) injury. Irreparable or irrecoverable injury is damage that is inextricably embedded in the cellular/subcellular fabric of the organism.[43] The Gompertz function becomes:

$$G_x = G_o + \alpha x + \epsilon \qquad (7)$$

where ϵ represents the increment of irreversible injury. It matters not if the injury is qualitatively dissimilar to aging injury; all that is required is that the added injury alters system states so as to enhance the probability of death. If exposure is continued at a zero-order rate at intervals of x units, and injury is accrued at an age-independent rate, the function becomes:

$$G_x = G_o + \alpha x + \gamma x \qquad (8)$$

where γ is a first-order toxicity rate constant. If the injury is instantaneous but repaired by a first-order process, we have:

$$G_x = G_o + \alpha x + \epsilon e^{-Kx} \qquad (9)$$

where K is the first-order rate constant specifying repair. If the injury is continuous (zero-order) but repaired at a first-order rate, the function is:

$$G_x = G_o + \alpha x + \frac{\lambda (1 - e^{-Kx})}{K} \qquad (10)$$

where λ is a hybridized parameter proportional to the zero-order rate constant specifying toxicity injury.

Hypothesis VI: Longevity hormesis is a response to a biologically nonessential, exogenous agent or stimulus that reversibly reduces Gompertzians relative to a control population without affecting senescence (i.e., G_o or ϕ).

This hypothesis, while positing the concept of longevity hormesis, does little more than empirically describe it. As mentioned previously, a definition must await characterization of mechanism. Three points are worth reiterating. First, longevity hormesis is reversible. Second, longevity hormesis is presumed to be mediated through a mechanism that does not impact on senescence, but rather superimposes its effects onto it. And third, longevity hormesis is only produced by nonessential stimuli. Application of life-enhancing essential agents (nutrients, vitamins, etc.) promote longevity through what has been termed a "proper" action.[5] Unfortunately, the distinction between "proper" and "longevity hormetic" enhancement of longevity is empiric, based solely on our knowledge of process and system.[47] While it is possible that biologically essential agents might also prompt a longevity hormetic response, there is no currently available data demonstrating this.

In terms of superimposition onto the Gompertz function, longevity hormesis is expressed by a term that reversibly reduces Gompertzians. By way of example, assume a linear Gompertz function characterizing senescent aging in concert with a longevity hormetic stimulus that reduces injury at a zero-order rate; the injury reducing effect, however, is reversible and dissipates at a first-order rate. The function is identical with Equation 10, except the far right term has a negative sign designating injury reduction:

$$G_x = G_o + \alpha x - \frac{\lambda (1 - e^{-Kx})}{K} \qquad (11)$$

Although this example denotes one kinetic scheme for input and dissipation of longevity hormesis, it is emphasized that any appropriate empiric function may be used. For example, Thompson et al.[47] developed a model of longevity hormesis in wild chipmunks following a single dose of ionizing radiation; they found that aging apparently promoted the loss of longevity hormesis. This is in contrast to Equation 11, in which dissipation of longevity hormesis is age-independent.

For a phenomenon to be considered longevity hormetic, reduction in Gompertzians must be characterized by a Gompertz term different from that observed with caloric restriction—the implication being that the mechanisms are also different. When a linear Gompertz function describes control animals, caloric restriction usually acts to reduce mortality simply by reducing α;[28] that is, it slows the rate of endogenous injury generation.

When longevity hormesis is observed, it invariably occurs without the weight reduction observed in calorically restricted laboratory animals.[21,24,25]

Hypothesis VII: For homogeneous laboratory populations of eutherian mammals housed under uniform, good laboratory conditions and kept free of preventable disease, exposures to nonbiological, nonessential, exogenous agents or stimuli will not slow senescent aging.

> Oh, come with old Khayyám, and leave the Wise
> To talk; one thing is certain, that Life flies;
> One thing is certain, and the Rest is Lies;
> The Flower that once has blown for ever dies.
>
> Rubáiyát of Omar Khayyám[29]

In the linear Gompertz function, senescent aging is expressed through the vulnerability parameter (G_o) and α. It is hypothesized that exogenous agents, if toxic, are without impact on these two terms per se; rather they superimpose their injury, reversibly or nonreversibly, onto aging injury.[14] If life prolongation is achieved through exposure to exogenous agents, this hypothesis indicates the only direct effect can be through longevity hormesis. Thus, pharmacological treatments, administration of antioxidants, stress, etc., are without impact on senescence. It appears that for eutherian mammals, housed in an optimum environment, the only way to reduce the rate of senescence is through dietary modification and possibly physical activity. Consistent with this view is the finding by Sacher[5] that the only factor affecting ϕ in eutherian mammals housed under ideal conditions is caloric restriction. Although Sacher did report that exogenous agents could affect the vulnerability parameter, he appears to have neglected the fact that at zero time (usually taken at or near weaning), both treated and untreated animals must have the same hazard function and Gompertz intercept. Thus, Sacher applied an inappropriate extrapolation to zero time.

If Hypothesis VII is true, then, contrary to popular notion, pharmacologic treatments such as antioxidants and selegiline (deprenyl) administration will not directly affect either of the two aging parameters of the linear Gompertz function in a life-enhancing fashion. Of course, it is always possible for pharmacologic treatments to affect dietary consumption and/or weight gain, thereby indirectly impacting on aging. This seems likely to have been the mechanism through which some antioxidants, in earlier reports, enhanced longevity in mammalian population studies; by virtue of toxicity and/or taste, these antioxidants probably acted by abating appetite, thus reducing caloric consumption in a manner favoring longevity.[16] This is consistent with Cutler's[32] conjecture that cellular antioxidants have overlapping capabilities, allowing one to replace another in a negative feedback loop. Thus, high-level dietary vitamin E supplementation, for example, would depress levels of other endogenous antioxidants, maintaining net antioxidant protection at a fairly constant level.

This hypothesis naturally leads to the question of how advantageous pharmacologic agents enhance longevity. For example, it is well established that administration of antihypertensive agents to individuals with high blood pressure reduces population mortality. Viewed in the context of this discussion, antihypertensive agents do not slow senescence, but rather modulate senescent symptoms in a positive, life-enhancing fashion. In other words, senescent toxicity is still present, but its maleficent manifestations, which in turn could elicit further injury and destruction (e.g., stroke, myocardial infarction, etc.), are temporarily held in abeyance. This concept also seems to apply to drugs that lower cholesterol in humans. The author's unpublished Gompertz analysis of the data from Anderson et al.[48] demonstrates that the population with elevated serum cholesterol had Gompertzians upwardly displaced in a near-parallel fashion relative to subjects with normal cholesterol. As discussed previously,[5,14] this indicates reversible toxicity. Serum cholesterol–lowering drugs presumably return the patient to a senescent state that he or she would be at in the absence of elevated serum cholesterol; that is, cholesterol lowering drugs modulate senescent symptoms, manifestations, and cascading effects, but not senescence itself.

The skeptic, justifiably, would take strong exception to this view. Ultimately, however, most disagreements come full circle, returning to the thorny issue of just what constitutes aging! As Huber Warner, chief of the molecular and cell biology branch at the U.S. National Institute on Aging put it, "If you ask what's normal aging, we still can't tell you."[49] Restated poetically in metaphor:

There was the Door to which I found no Key:
There was the Veil through which I could not see:
Some little talk awhile of Me and Thee
There was — and then no more of Thee and Me.

Rubáiyát of Omar Khayyám[29]

The approach taken here, admittedly arbitrary, is that biological disorder, cell dysdifferentiation, variations in constraints, etc., produce aging, and that disease is a manifestation of aging, not aging itself.

Hypothesis VIII: Relative to other eutherian mammalian species, humans' enhanced longevity (e.g., maximum life span potential) is due to a neotenous slowing of orthodox mammalian aging, as opposed to a qualitatively distinct aging process.

Applying allometric scaling principles to the entire class *Mammalia*,[50-53] and assuming a body mass of 70 kg for humans, our species should have a gestation time of 6.08 months, a maximum life span of 26.9 years, and a brain mass of 269 g. Realistic values of these variables are 9.0 months, 114 years, and 1400 g, respectively.[50,54] The sizable deviation of these human variables vis-à-vis the allometric regression line has been termed *vertical allometry*.[51]

These characteristics, as well as such traits as large brain size, erect posture, small teeth, thin nails, prolonged growth period, late development of sexual maturity, ventral positioning of the vagina (permitting us to express our sexuality facing one another), relative hairlessness, etc., all result from a phenomenon termed *neoteny*.[55] As defined by Gould,[52] neoteny is the "retention of formerly juvenile characters by adult descendants . . . produced by retardation of somatic development." As it specifically relates to humans, it is "the preservation in adults of shapes and growth rates that characterize juvenile stages of ancestral primates"[56] (see Figure 1.1 for an illustration[57]). As a result of neoteny, we are born at an earlier stage of physical development than other primates and do not mature as far along their developmental path.[55,58] The retardation feature of neoteny is believed to have provided the only "escape" from our ancestral allometry, permitting our species to embark evolutionarily toward a favored adaptation.[52] With respect to our aging, it is precisely our neotenous nature that enhances our longevity relative to other mammalian species of similar size.

Figure 1.1.   Baby and adult chimpanzees. Note the differences in facial morphology as a function of age, but also the strong resemblance between the baby chimp and adult humans. Naef[57] commented: "Of all the animal pictures known to me, this [baby chimpanzee] is the most man-like." As a result of neoteny, we humans develop more slowly than other mammalian species and, as adults, retain many juvenile features of our phylogenetic ancestors. While we did not descend from chimpanzees, the juveniles of our phylogenetic ancestors probably looked similar to the baby chimp illustrated here. This figure, and one similar to it, were previously used by Gould[52] and Montagu[55] to illustrate the same point. See Montagu[55] for an excellent monograph on neoteny. Reprinted from Naef[57] by permission of Springer-Verlag Publishers.

Interestingly, the concept of neotenous development in primates parallels the view that the human infant is born "premature," having completed only about half its gestation period inside the womb (uterogestation), with another 10 months required outside the womb (exterogestation).[55] The relative helplessness of the exterogestation period can be presumed to end with quadrupedal locomotion. Birth is simply a bridge between intra- and extrauterine gestation. Human babies are born "prematurely" for two reasons—large fetal brains and limitations on female pelvic size. It has been speculated that the incongruities of exterogestation (omnipotent pleasure indulgence vs powerless dependence on others) sow the seeds of human conflict and neurosis.[59]

Scientifically, animals are utilized as biological models (surrogates) of humans primarily because they function well at four levels of interest, that is, at four different time scales:[60-62]

1. physicochemical (microseconds to minutes)
2. physiological (minutes to days)
3. morphogenetic (days to decades)
4. evolutionary (several lifetimes)

Because virtually all living things are made from cells, are based on the same genetic code, and evolved by natural selection, and because all life is connected,[63] it would appear that all living things possess common controlling elements capable of imposing fundamental and universal properties, including aging.

Hypothesis IX: In eutherian mammalian species, exposures to particular, nonessential, exogenous agents or stimuli capable of stressing the system, produce a degree of instability that triggers a homeostatic, longevity hormetic response, temporarily (reversibly) reducing the sum total of injury within the organism.

All living systems tend to maintain homeostasis, keeping an orderly balance among subsystems that process matter-energy or information.[17] They do so principally by employing pattern recognition and negative feedback; pattern recognition extracts crucial information amidst a flood of irrelevant signals, whereupon negative feedback acts on this information, returning the system to near its original state. A system state is pathological (1) when one or more of its variables perseveres beyond its range of stability for any significant period of time, or (2) when the costs of adjusting an ailing system are significantly increased. How efficiently a system adjusts is determined by what strategies are employed and whether they satisfactorily reduce strains without being too costly. In the case of most noxious stimuli exposure, the system cannot remove or circumvent them; consequently, its only course of action is to master them in a cost-efficient fashion. Longevity hormesis appears to be such a response; its reversibility serves to keep it economical.

Hypothesis X: When, through exposure to nonessential, exogenous agents or stimuli, longevity hormesis is elicited in homogeneous laboratory populations of eutherian mammals housed under uniform, good laboratory conditions and kept free of preventable disease, toxicity, if also induced, is most likely to be of the irreversible kind.

In searching the literature for mortality data in laboratory populations, the most frequently located data come from studies in which exogenous agents are administered at a constant rate. When toxicity is observed, with or without concomitant longevity hormesis, it most frequently is the irreversible kind; longevity hormesis, by definition, is reversible. Assuming both irreversible toxicity and reversible hormesis are deposited onto the linear Gompertz function at a zero-order rate, and further assuming longevity hormesis dissipates at a first-order rate, one obtains the following Gompertz function:[18,24,25]

$$G_x = G_o + \alpha x + \gamma x - \frac{\lambda (1 - e^{-Kx})}{K} \tag{12}$$

where γx represents the irreversible accumulation of toxicity injury and the far right term reflects the reversible longevity hormetic effect.

Obviously both γ and λ will be dependent on the dose of exogenous agent. Neafsey et al.[24,25] empirically found that the logarithmic-logistic function,[64-74] also known as the *sigmoid E_{max} model, generalized hyperbolic function,* and *Hill equation,* characterized the dose-response relationships remarkably well. This is not to imply that toxicity and longevity hormesis processes obey the underlying assumptions inherent in this equation, but rather that the large number of parameters in the logarithmic-logistic equation provide superior flexibility in curve-fitting equations to data. It is therefore not surprising that in the examples cited below, Equation 12 (or a "collapsed" version) is most frequently employed.

In generalized nomenclature, the logarithmic-logistic equation may be expressed as follows:

$$R = \frac{R_m D^s}{(1/Q) + D^s} = \frac{R_m D^s}{(ED_{50})^s + D^s} \tag{13}$$

$$ED_{50} = Q^{-(1/s)} \tag{14}$$

where   R = response
          R_m = maximum response
          D = dose

$$Q = \text{a ``scale'' or ``affinity'' parameter}$$
$$s = \text{a ``shape'' parameter}$$
$$ED_{50} = \text{the dose producing one-half of } R_m$$

An example of "collapsing" occurs when $D^s >> (1/Q)$; under these conditions, R approximates R_m. In toxicity studies, where doses are relatively large, it is possible to saturate the longevity hormetic response at the lowest dose; this occurred with γ-radiation exposure to male and female mice (lowest dose: 0.11 rad/day), as well as with hexachlorobenzene feeding to female rats (lowest dose: 0.32 ppm in diet).[25] As noted previously,[7] longevity hormesis is frequently a high-affinity, low-capacity phenomenon; that is, it is manifested and reaches its maximum effect at relatively low doses. Toxicity, on the other hand, frequently acts more like a low affinity, high capacity phenomenon; that is, it is only manifested at "high" doses, and there is virtually no limit (excepting death) to the damage it can inflict.

THE LONGEVITY HORMESIS DATABASE

Because of (1) the loose use of the term *(longevity) hormesis*, (2) the ease with which longevity enhancement from caloric reduction can be confused with that from longevity hormesis, and (3) the need to analyze age-specific mortality rate data using appropriate models and weighting factors, only those data sets analyzed within the framework specified by Neafsey et al.,[24,25] and meeting the appropriate criteria cited therein, will be considered authenticated. Authentication, however, should only be construed as implying consistency, as opposed to validation (only when the mechanism becomes known and can be experimentally established will validation be possible). The criteria for consistency were:

1. a good randomness of scatter of data about the fitted curves (judged by weighted residual plots and visual inspection of the curve-fits[24,25,75]
2. the computed chi-square values were less than the tabulated value ($\alpha = 0.05$)[14,24,25,41]

It should be noted that the data analyzed by Neafsey et al.[24,25] focused exclusively on systems in which mortality from control populations could be characterized by a linear Gompertz function; because the purpose of these two papers by Neafsey et al.[24,25] was to lay a foundation for the longevity hormesis concept vis-à-vis Gompertz mortality analysis, it was decided at that time to simplify the database as much as reasonably possible.

In addition to data sets meeting the specific criteria, there appear to exist other data strongly suggesting the presence of longevity hormesis.[5,14,24,25,33,47,76-78] These data will also be collated. Table 1.1 summarizes the database.

Table 1.1. Longevity Hormesis Database

Stimulus	Species (Gender)	Dose (Route)	Concomitant Toxicity?	Reference(s)
		Authenticated Data		
Procaine	rats (male)	4 mg/kg 3 × weekly (parenteral)	no	5,14, this chapter
Amosite asbestos[a]	rats (female)	10,000 ppm (diet)	yes	24
Amosite asbestos	hamsters (female)	10,000 ppm (diet)	yes	24
Amosite asbestos	hamsters (male)	10,000 ppm (diet)	yes	24
Dieldrin[a]	mice (male)	1 ppm (diet)	yes	24
Ethyl acrylate	rats (male)	75 ppm (inhalation)	yes	24
Methylene chloride	hamsters (female)	500–3500 ppm (inhalation)	no	24,25
Chloroform[a]	rats (male)	1800 ppm (water)	yes	24
Gamma radiation	mice (mixed)	0.11–8.8 rad/day (whole body)	yes	25,33
Gamma radiation	mice (male)	0.11 rad/day (whole body)	no	25
Gamma radiation	chipmunks (male & female)	200–400 R single-dose (whole body)	yes	47
Hexachlorobenzene[a]	rats (female)	0.32–40 ppm (diet)	yes	25
DDT[a]	mice (female)	2–250 ppm (diet)	yes	25
DDT[a]	mice (male)	2–250 ppm (diet)	yes	25
		Gompertz Plots Strongly Suggesting Longevity Hormesis		
2-Mercaptoethanol	mice (male)	0.25% (w/w) (diet)	no	76
Crowding conditions	rats (male)	6 vs 12 rats per cage	no	77
X-radiation	*Drosophila* (male)	1–20 kR (whole body)	yes	33,78

[a]Visual inspection of the Gompertz plots, while indicating consistency with the longevity hormesis/toxicity model posited, leaves room for other interpretations. This is due to (1) a weak longevity hormetic effect; (2) cancellation of the reverse effects of longevity hormesis by irreversible toxicity; and/or (3) variability in the data.

LONGEVITY HORMESIS: DISCUSSION OF THE AUTHENTICATED DATA

Why then should I marvel or let myself be frightened because one part is poison, and despise the other part too? . . . Now if the poison conquers not but enters without harm when we use it according to nature's ordered way, why then should poison be despised? Who despises poison, knows not what is in the poison. . . . If you wish justly to explain each poison, what is there that is not poison? All things are poison, and nothing is without poison: the *Dosis* alone makes a thing not poison.

Paracelsus[79]
(1493–1541)

Before discussing actual data, it would be helpful to examine simulations based on an empirically useful Gompertz function, Equation 12. Superimposed on a linear Gompertz function, this equation posits:

1. a zero-order input of irreversible injury
2. a zero-order input of longevity hormesis
3. a first-order dissipation of retained longevity hormesis

To eliminate either toxicity or longevity hormesis, γ or λ is set equal to zero, respectively. To account for dose-dependent effects, γ and λ may be expressed as a function of dose (e.g., by utilizing the logarithmic-logistic equation). Figure 1.2 illustrates a few of the possible curves that can be

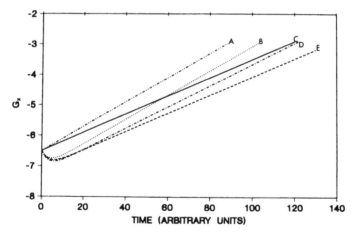

Figure 1.2. Gompertz diagrams illustrating the effects of irreversible toxicity and/or longevity hormesis on the linear Gompertz function. Curve C represents a control population whose mortality experience is characterized by a linear Gompertz function. Curve A assumes superimposition of only irreversible toxicity; curve E assumes superimposition of only reversible longevity hormesis; and curves B and D assume both irreversible toxicity and reversible longevity hormesis simultaneously superimpose their effects on the control function (toxicity is more dominant in B, i.e., a larger γ). Reprinted from Neafsey et al.,[24] p. 376, by permission of Marcel Dekker, Inc.

generated in this fashion. Curve C shows a linear Gompertz function. Superimposing irreversible toxicity upon Curve C, one obtains Curve A. Note that the intercept remains the same but that the slope becomes steeper. The impact of longevity hormesis is illustrated by Curve E. With zero-order input and first-order elimination, longevity hormesis quickly reaches a steady state. When this occurs, curves C and E become parallel to one another. The condition of having both irreversible toxicity and longevity hormesis superimposed upon the linear Gompertz function is illustrated with Curves B and D. Toxicity (expressed through γ) is greater in B than D.

A good example of a system in which pure longevity hormesis is superimposed on a linear Gompertz function is illustrated in Figure 1.3 for procaine administration to male rats. Interestingly, this response is not observed in female rats. This indicates, as do some other data sets, that longevity hor-

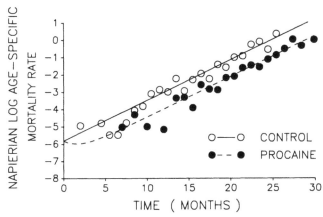

Figure 1.3. Gompertz plots for untreated male rats (control) and those receiving procaine: a classic longevity hormesis response in the absence of concomitant toxicity (analogous to curve E, Figure 1.2). Animals used were white rats of the French Wistar strain. Treated animals received procaine "parenterally" at a dose of 4 mg/kg three times weekly for 4 weeks, whereupon treatment was discontinued for 1 month. Beginning at either 2 or 6 months of age, the injections were continued for the remainder of each animal's life span. The control population received saline injections. As the age of initiation of uninterrupted therapy (2 or 6 months) did not apparently affect mortality, data from these two groups were pooled. Time on the abscissa is equivalent to age. The linear Gompertz function (Equation 3) was used to characterize mortality experience for the control group. Equation 11, which superimposes a longevity hormesis term onto the linear Gompertz function, was used to characterize the procaine-treated population. Both equations were fit (least squares analysis) simultaneously to the data using appropriate weighting factors (see Neafsey et al.[24] for methodology). Parameter estimates were G_o = −5.810, α = 0.2323 hr^{-1}, λ = 0.4953 hr^{-1}, and K = 0.5095 hr^{-1}. These data, published by Aslan et al.,[26] were previously analyzed by Sacher[5] and Boxenbaum et al.;[14] in both analyses, independent linear regressions indicated that the line from the procaine-treated group was displaced downward in a parallel fashion from that of controls. No longevity hormetic effect was evident from data in female rats.

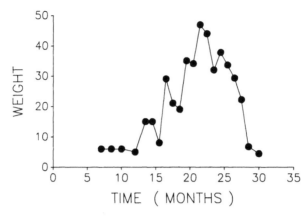

Figure 1.4. Statistical "weights" vs time data used for the procaine-treated group in the nonlinear least-squares analysis illustrated in Figure 1.3. Weights (estimates of the reciprocal of variance) were calculated from Equation 16. See Sacher[2] and Boxenbaum et al.[14] for further discussions of Gompertz plot weighting. In unweighted least-squares analysis, one attempts to minimize the sum of squared deviations by parameter adjustment (iteration). In weighted least-squares analysis, it is the weighted sum of squared deviations that one attempts to minimize. The weighting procedure is used to adjust for the magnitude and "clout" of each dependent variable. See Boxenbaum et al.[75] and Daniel and Wood[80] for general discussions.

mesis can be gender specific. Figure 1.4 illustrates statistical weights used in the regression analysis for the procaine treated population. In conventional (unweighted) least squares analysis, it is assumed (usually tacitly) that variance estimates about each data point are equal.[80] However, in Gompertz analyses, particularly those employing cohorts, this is hardly ever the case. Therefore, it is best to weight the dependent variable in accordance with an estimate of the reciprocal of its variance.[81] Taking cognizance of this, Sacher[2] developed relationships to estimate the sampling variance:

$$P_i = (N_i - d_i)/N_i \qquad (15)$$

$$w = 1/v \cong [N_i\, P_i\, (\ln p_i)^2]/[1 - P_i] \qquad (16)$$

where N_i = the number of survivors at the beginning of an age interval
        d_i = the number of deaths over that same interval
        w = the statistical weight
        v = variance

Time is taken as the midtime of the interval (as is Ω_x).

For studies in which survival data are not reported but mean and/or median survival times are, it is not possible to unequivocally explain observations of longevity enhancement. Consider Figure 1.5, which illustrates the impact of caloric restriction on linear Gompertz plots of male laboratory rats.[28] Note that α is reduced in the calorically restricted animals, without a change in ln Ω_o. This alteration of the linear Gompertz function is distinctly different from that observed with longevity hormesis. Yet, both caloric restriction and longevity hormesis enhance mean and median survival times. It is therefore recommended that investigators not only report mean and/or median survival times, but also provide survival/mortality data (preferably in the form of Gompertz plots).

In order to illustrate how longevity hormesis and caloric restriction affect the survivalship functions of linear Gompertz models, Figures 1.6 and 1.7 were constructed. In both figures, the control curves are identical, although scales on the abscissa differ. For the linear Gompertz model, the survivalship function is the following:[41]

$$S(x) = \exp -[(h(x)_o/\alpha)\,(e^{\alpha x} - 1)] \qquad (17)$$

where $S(x)$ is fractional survival at time x, $h(x)_o$ is the hazard function at zero-time, and the other terms have already been defined. Although this equation was not explicitly used to construct the control animal survivalship

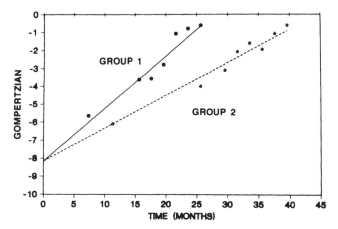

Figure 1.5. Gompertz plots illustrating the effect of caloric restriction on the Gompertz function of laboratory rats (male Fischer 344 strain). The study commenced utilizing 26- to 30-day old weanling rats. Time on the abscissa began at 6 weeks of age. The group 1 population consisted of 40 male animals who were fed *ad libitum*. The group 2 population consisted of 40 male animals who were fed at about 60% the mean caloric intake of population 1 until 18 months of age, and then maintained at their 18 month caloric intake until death. To adjust for vitamin and mineral intake, the diet of group 2 animals contained 1.53-fold more minerals and 1.66-fold more vitamins. The linear Gompertz function (Equation 3) was fit to all data simultaneously (using different values of α for different feeding regimens), employing weighted least-squares regression analysis. For calorically restricted animals, note the reduction in α (slope), with no alteration in the vulnerability parameter (intercept). The original data came from Yu et al.[82] Reprinted from Neafsey et al.,[28] p. 360, by permission of Marcel Dekker, Inc.

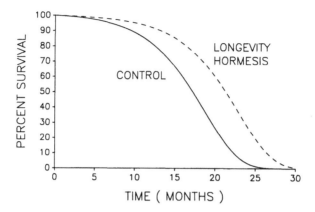

Figure 1.6. Theoretical survival curves for a control population and one in which longevity hormesis is operative. The survival equation for the control group corresponds to a linear Gompertz function (Equation 3); Gompertzians for the longevity hormesis group are characterized by Equation 11. Parameters are identical to those in Figure 1.3. Survival percentages were calculated using Equations A-2 and A-4 of Neafsey et al.[24] Note that at approximately 50% survival, the two curves appear somewhat parallel.

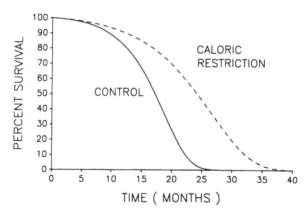

Figure 1.7. Theoretical survival curves (linear Gompertz model) for a control population and one in which caloric restriction reduces α. The survival function for the control population is characterized by values of α and the vulnerability parameter equal to those for the control population in Figures 1.3 and 1.6. The calorically restricted group was assigned a value of α equal to 63.0% of controls, since this corresponds to the reduction in α noted for caloric restriction by Neafsey et al.[28] and illustrated in Figure 1.5. Survival percentages were calculated using Equations A-2 and A-4 of Neafsey et al.[24] Note that at approximately 50% survival, the curve for the calorically restricted group appears to have a less steep slope than that for controls. Contrast this with Figure 1.6, where longevity hormesis is the factor enhancing longevity.

curves in Figures 1.6 and 1.7, an equivalent method was used (Equations A-2 and A-4 of Neafsey et al.[24]). In comparing Figures 1.6 and 1.7, note the differences in the curves for calorically restricted animals relative to those experiencing longevity hormesis; they are distinctly different in relationship to the control group.

Figure 1.8 provides another example of a system (male LAF1 rats receiving 0.11 rad/day γ-radiation) in which longevity hormesis exists in the absence of observable toxicity.[25] This is in contrast to Figure 1.9, which illustrates the effects of several doses of γ-radiation on Gompertz plots.[25] At the 0.11rad/day dose, there is both minor longevity hormesis and toxicity; however, in this study, data from both genders were pooled, whereas in Figure 1.8 only data from male animals are illustrated. Note in Figure 1.9 how increased doses of radiation induce toxicity to the point that longevity hormesis is obscured. This is a major problem in long-term toxicity studies, which typically employ relatively high doses of toxicants (generally about 12.5 to 100% of the maximum tolerated dose per day) in an attempt to assess risk at much lower doses.

Although both Boxenbaum et al.[18] and Neafsey[27] have recently addressed the problem of potentially overlooked longevity hormesis, the risk assessment community has failed to give it serious consideration. Previously, Smyth[85] had taken notice of the fact that low doses of otherwise toxic substances can be beneficial. His reward: the epithet "Dr. Smyth and his fellow poisoners."[86] Although the scientific community envisages itself as the epitome of institutionalized rationality,[87] many researchers have noted

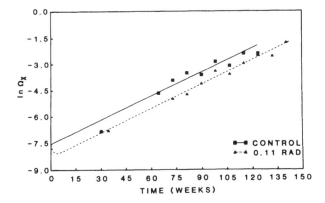

Figure 1.8.   Gompertz plots for control and γ-radiation treated male LAF1 mice. The treated group received γ-radiation at a dose of 0.11 rad/day (administered over 8 hr), begun at 1 month of age and continued throughout life. Time on the abscissa refers to the period following initiation of exposure. Note the classic longevity hormesis pattern, with no apparent concomitant toxicity. The theoretical lines were obtained by simultaneous fitting of Equations 3 and 11, employing weighted least-squares regression analysis. The original data came from Lorenz et al.[83] Reprinted from Neafsey et al.,[25] p. 140, by permission of Marcel Dekker, Inc.

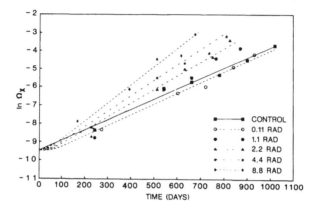

Figure 1.9. Gompertz plots for control and γ-radiation treated male and female (pooled) LAF1 mice. Exposure to γ-radiation (0.11–8.8 rad/day administered over 8 hr) was begun at approximately 70 days of age and subsequently continued throughout life. Time on the abscissa refers to the period following initiation of exposure. The theoretical lines were obtained by simultaneous fitting of Equations 3 and 12 (denoting both longevity hormesis and irreversible toxicity), employing weighted least-squares regression analysis. In the mathematical model, maximum longevity hormesis appeared to have been reached at the lowest dose (0.11 rad/day). The toxicity parameter, γ, was characterized by a logarithmic-logistic function. Note that the influence of longevity hormesis only dominates (meagerly) with the 0.11-rad/day group, whereas the impact of irreversible toxicity is principally observed at the other dose levels. The original data came from Lorenz et al.[84] Reprinted from Neafsey et al.,[25] p. 138, by permission of Marcel Dekker, Inc.

the high degree to which anomalous information is ignored if it disconfirms basic assumptions of established paradigms.[88] Once a group agrees that a particular kind of reality is desirable, they develop a style that permits them to deal with observations solely on their own terms — and woe to the individual with different ideas.[89] For most individuals, escape from these intellectual-scientific fetters is difficult, for the obduracy of established perspective locks practitioners together in a rigid framework of beliefs that is not readily overcome.[88,90]

Further support that γ-radiation produces longevity hormesis is supplied in Figures 1.10 and 1.11. However, in this case, the data deal with chipmunks living in the wild. The animals were live-trapped, irradiated with either a single dose of 200 or 400 R γ-radiation (except for controls), and then returned to the wild. The original publication[47] equated Roentgens with rads, although more specifically 1 R is approximately 0.95 rad with respect to muscle tissue.[91] Figure 1.10 illustrates representative survival curves. It is readily apparent that γ-radiation exposure, within the dose range utilized, enhanced longevity. Figure 1.11 illustrates differences in Napierian logarithms of age-specific mortality rates between treated and control animal populations; an arbitrary value of 5 was added to ensure positive numbers throughout. Values below 5 indicate net longevity horme-

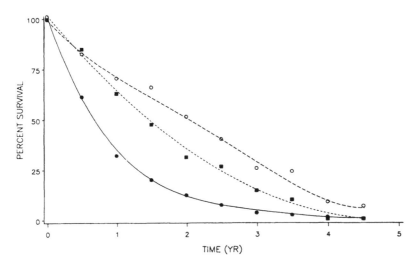

Figure 1.10. Representative survivalship functions for eastern chipmunks *(Tamias striatus)* living under natural conditions in the wild. These curves characterize data for female animals of known birthdate at location site 101 in Crawford County, Pennsylvania. Empiric polynomial functions were fitted to the three sets of population data: ● = controls; ■ = animals receiving a single dose of 200 R ionizing radiation; and ○ = animals receiving a single dose of 400 R. Time on the abscissa began at capture. Note the improved survival of irradiated animals. Thompson et al.[47] equated Roentgens with rads; however, a more appropriate conversion factor is 1 R is approximately 0.95 rad with respect to muscle tissue.[91] The original data came from Tryon and Snyder.[92] Reprinted from Thompson et al.,[47] p. 276, by permission of Marcel Dekker, Inc.

sis; values above 5 indicate net toxicity. Initially, longevity hormesis dominates, but it dissipates over time. By the end of the study there was relatively minor toxicity. In the work from which the original chipmunk data were obtained,[92] the investigators also administered 900–1700 R to 16 captive chipmunks. Within 30 days, all but one died. So, whereas 400 R is beneficial to longevity, 900–1700 R is highly toxic. By way of comparison, the LD_{50} for γ-radiation in female laboratory mice (RF strain) is approximately 736–1053 rad.[93]

Figure 1.12 illustrates another example of a system (employing methylene chloride exposure to female hamsters) in which longevity hormesis, in the absence of toxicity, is produced.[25,94] A logarithmic-logistic equation was used to characterize λ, and consequently, the larger the dose, the greater the longevity hormesis. In stark contrast, Figure 1.13 illustrates the effect of methylene chloride exposure (same ppm in the air) on the Gompertz plots of female rats.[25,94] No longevity hormesis here, only irreversible toxicity. This time a logarithmic-logistic equation was used to characterize irreversible toxicity. These two figures exemplify one of the major problems in toxicology and risk assessment — interspecies extrapolation.

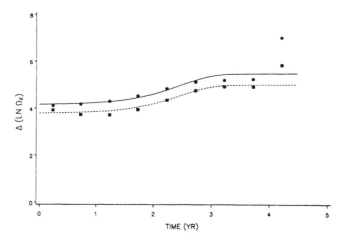

Figure 1.11. Graphical representation of the longevity hormesis–toxicity Gompertz model for eastern chipmunks, living in the wild, exposed to a single dose of either 200 or 400 R γ-radiation. These curves were derived from the same data set illustrated in Figure 1.10. Time on the abscissa began at capture. Values of Δ (ln Ω_x) denote differences in Napierian logarithms of age-specific mortality rates between treated and control animal populations; an arbitrary value of 5 was added to ensure positive values throughout (for curve-fitting purposes). Therefore, values below 5 indicate net longevity hormesis, whereas those above 5 indicate net toxicity. Filled squares and filled circles represent Δ (ln Ω_x) data for the 200- and 400-R groups, respectively. The theoretical curves were determined by fitting appropriate equations to the data, employing weighted nonlinear least-squares analysis. The original data came from Tryon and Snyder.[92] Reprinted from Thompson et al.,[47] p. 279, by permission of Marcel Dekker, Inc.

Figure 1.14 illustrates the effect of amosite asbestos administration on the Gompertz plot of exposed female rats.[24,95] Although a mixed longevity hormesis–irreversible toxicity model is consistent with the data, the weak longevity hormetic effect, coupled with the apparent cancellation of one effect by the other, makes model-independent interpretation difficult. This, however, is not true for amosite administration to female hamsters at the same exposure. Figure 1.15 clearly illustrates the effects of both longevity hormesis and irreversible toxicity.[24,96] Gompertz plots for male hamsters follow the same pattern as female hamsters, as illustrated in Figure 1.16.

Figure 1.17 provides another example where visual inspection of Gompertz plots leaves the viewer wondering just how likely the longevity hormesis–irreversible toxicity model really is.[24,97] It illustrates plots for control and dieldrin treated male mice. Given data variability and potential cancellation of effects, one is tempted to merge the data, drawing a single straight line through all data points.

Figure 1.18 is a good example of another problem. It illustrates the influence of ethyl acrylate inhalation (75 ppm) on the Gompertz plot for male rats.[24,98] The case for the existence of longevity hormesis would seem to

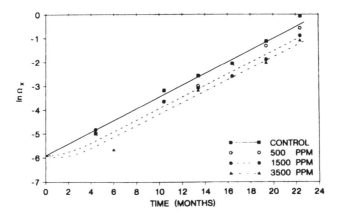

Figure 1.12. Gompertz plots for control and methylene chloride–treated female Syrian golden hamsters. Methylene chloride exposure (500–3500 ppm inhalation) was begun at 8 weeks of age and continued for 2 additional years (6 hours per day, 5 days per week). Time on the abscissa refers to the period following initiation of exposure. The theoretical lines were obtained by simultaneous fitting of Equations 3 and 11, employing weighted least-squares regression analysis. The logarithmic-logistic equation was used to characterize λ. Note a dose-dependent increase in longevity hormesis with increase in dose, in the absence of any apparent toxicity. The original data came from Burek et al.[94] Reprinted from Neafsey et al.,[25] p. 134, by permission of Marcel Dekker, Inc.

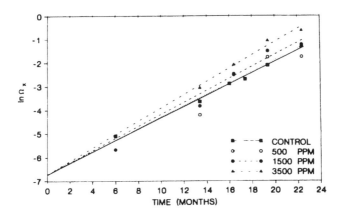

Figure 1.13. Gompertz plots for control and methylene chloride–treated female Sprague-Dawley rats. Methylene chloride exposure (500–3500 ppm inhalation) was begun at 8 weeks of age and continued for 2 additional years (6 hours per day, 5 days per week). Time on the abscissa refers to the period following initiation of exposure. The theoretical lines were obtained by simultaneous fitting of Equations 3 and 8, employing weighted least-squares regression analysis. The logarithmic-logistic equation was used to characterize γ. Note the dose-dependent increases in irreversible toxicity, in the absence of any apparent longevity hormesis—just the opposite of what occurred in hamsters at the same exposure levels (see Figure 1.12). The original data came from Burek et al.[94] Reprinted from Neafsey et al.,[25] p. 136, by permission of Marcel Dekker, Inc.

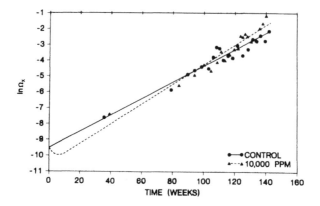

Figure 1.14. Gompertz plots for control and amosite asbestos–treated female F344/N rats. Amosite asbestos administration (10,000 ppm feed) was begun at 8 weeks of age and continued throughout life. Time on the abscissa refers to the period following initiation of exposure. Although the data are consistent with concomitant longevity hormesis and toxicity, and this model was used in the curve-fits, the figure is equivocal in making a more definitive judgment. The theoretical lines were obtained by simultaneous fitting of Equations 3 and 12, employing weighted least-squares regression analysis. The original data came from the National Toxicology Program.[95] Reprinted from Neafsey et al.,[24] p. 386, by permission of Marcel Dekker, Inc.

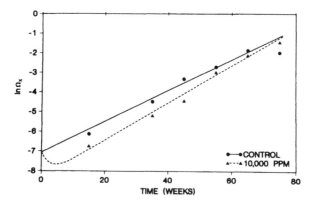

Figure 1.15. Gompertz plots for control and amosite asbestos–treated female Syrian golden hamsters. Amosite asbestos administration (10,000 ppm feed) was begun at 10 weeks of age and continued throughout life. Time on the abscissa refers to the period following initiation of exposure. The theoretical lines were obtained by simultaneous fitting of Equations 3 and 12, employing weighted least-squares regression analysis. Both longevity hormesis and irreversible toxicity are evident. The original data came from the National Toxicology Program.[96] Reprinted from Neafsey et al.,[24] p. 387, by permission of Marcel Dekker, Inc.

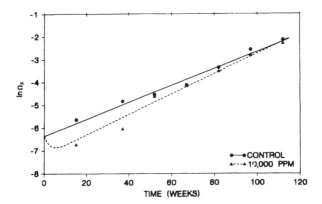

Figure 1.16. Gompertz plots for control and amosite asbestos–treated male Syrian golden hamsters. Amosite asbestos administration (10,000 ppm feed) was begun at 10 weeks of age and continued throughout life. Time on the abscissa refers to the period following initiation of exposure. The theoretical lines were obtained by simultaneous fitting of Equations 3 and 12, employing weighted least-squares regression analysis. Both longevity hormesis and irreversible toxicity are evident. The original data came from the National Toxicology Program.[96] Reprinted from Neafsey et al.,[24] p. 388, by permission of Marcel Dekker, Inc.

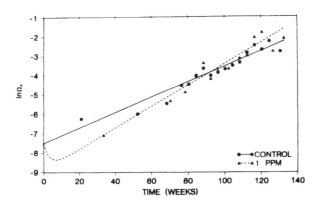

Figure 1.17. Gompertz plots for control and dieldrin-treated male CF-1 mice. Dieldrin administration (1 ppm feed) was begun at 4 weeks of age and continued throughout life. Time on the abscissa refers to the period following initiation of exposure. Although the data are consistent with concomitant longevity hormesis and toxicity, and this model was used in the curve-fits, the figure is equivocal in making a more definitive judgment. The theoretical lines were obtained by simultaneous fitting of Equations 3 and 12, employing weighted least-squares regression analysis. The original data came from Walker et al.[97] Reprinted from Neafsey et al.,[24] p. 389, by permission of Marcel Dekker, Inc.

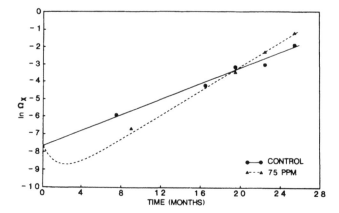

Figure 1.18.   Gompertz plots for control and ethyl acrylate–treated male F344 rats. Ethyl acrylate administration (75 ppm inhalation) was begun at 7–9 weeks of age and continued to 24 months of age (6 hr per day, 5 days per week). Time on the abscissa refers to the period following initiation of exposure. The theoretical lines were obtained by simultaneous fitting of Equations 3 and 12, employing weighted least-squares regression analysis. Both longevity hormesis and irreversible toxicity seem evident (see text for caveat). The original data came from Miller et al.[98] Reprinted from Neafsey et al.,[24] p. 390, by permission of Marcel Dekker, Inc.

reside with one data point, lying at 9 months and carrying the lowest weight of the four data points on the curve. Yet the fit is excellent, and the mixed longevity hormesis-irreversible toxicity model is consistent with the data. Obviously, additional studies need be conducted before a more definitive statement can be made concerning the likelihood of longevity hormesis being elicited by ethyl acrylate in male rats.

The mortality data illustrated in Figure 1.19 presented a unique opportunity to incorporate not only the effects of longevity hormesis and irreversible toxicity, but also the change in mortality pattern induced by caloric restriction.[24,99] Chloroform was administered to male rats in their drinking water at a concentration of 1800 ppm. This apparently caused them to consume less food and water. Consequently, a second control group was employed that attempted to mimic food and water consumption of the chloroform exposed population. The top curve of Figure 1.19 illustrates a linear Gompertz function for the conventional control population (i.e., the group permitted ad lib access to food and water). The second control group, not exposed to chloroform but having reduced food and water intake, was also characterized by a linear Gompertz function. However, as expected (see Figure 1.5), it was necessary to reduce α. The chloroform exposed group, on the other hand, apparently experienced three effects:

1. longevity hormesis
2. irreversible toxicity
3. a reduced α due to reduced caloric intake

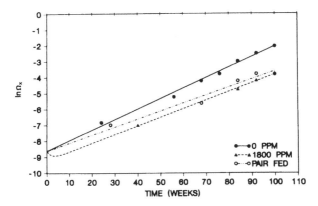

Figure 1.19. Gompertz plots for control and chloroform-treated male Osborne-Mendel rats. The curves illustrated are for three groups of animals: (1) control animals provided food and water *ad libitum*—no chloroform; (2) animals exposed to 1800 ppm of chloroform (drinking water) beginning at 7 weeks of age and continuing an additional 104 weeks; and (3) pair-fed controls attempting to match the reduced food and water intake of the chloroform-treated group, but without chloroform. Time on the abscissa refers to the period following initiation of exposure. For the first group, a linear Gompertz function was used to characterize mortality experience. For the third group (reduced food and water—no chloroform), a linear Gompertz function was also used, but with a reduced value for α (consistent with caloric restriction). For the second (chloroform) group, which had reduced caloric consumption and apparent longevity hormesis in concert with irreversible toxicity, the linear Gompertz function (with reduced α) was used together with terms for longevity hormesis and irreversible toxicity. All functions were fit simultaneously employing weighted least-squares analysis. The original data came from Jorgenson et al.[99] Reprinted from Neafsey et al.,[24] p. 395, by permission of Marcel Dekker, Inc.

While this model is most certainly non-unique, it is consistent and does demonstrate the potential of the generalized Gompertz model approach advocated herein.

Figures 1.20–1.22 illustrate Gompertz plots of the effects of hexachlorobenzene (Figure 1.20) and DDT (Figures 1.21 and 1.22) on rodent mortality.[25,100,101] In all three cases, the results are consistent with the longevity hormesis–irreversible toxicity model, although visual inspection suggests alternative models would also suffice.

PROBLEMS IN ASSESSING LONGEVITY HORMESIS IN HUMANS

As concluded by Neafsey,[21] evidence for the existence of longevity hormesis in humans is fragmentary at best. However, focusing on mortality statistics, there have been a few reports that suggest the possibility and/or potential of its occurrence in humans. Matanoski et al.[102] studied the mortality experience of radiologists (radiation exposure) relative to groups of three

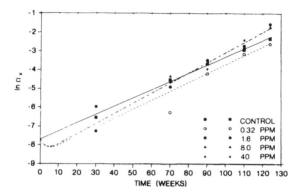

Figure 1.20. Gompertz plots for control and hexachlorobenzene-treated female Sprague-Dawley rats. Hexachlorobenzene administration (0.32–40 ppm feed) was begun at about 30 days of age and continued throughout life. Time on the abscissa refers to that period following initiation of dosing. Although the data are consistent with concomitant longevity hormesis and toxicity, and this model was used in the curve-fits, the figure is equivocal in making a more definitive judgment. The theoretical lines were obtained by simultaneous fitting of Equations 3 and 12, employing weighted least-squares regression analysis. In the mathematical model, maximum longevity hormesis appeared to have been reached at the lowest 0.32-ppm dose. The toxicity parameter, γ, was characterized by a logarithmic-logistic function. The original data came from Arnold et al.[100] Reprinted from Neafsey et al.,[25] p. 141, by permission of Marcel Dekker, Inc.

other physician specialists. For those radiologists who joined the Radiological Society of North America after the year 1940, mortality rates were initially lower than all other physician specialties; however, with follow-up 15–20 years later, mortality increased to values greater than ophthalmologists. This is akin to the longevity hormesis-irreversible toxicity model previously discussed for animal populations (Equation 12) and illustrated for γ-radiation in male and female mice (Figure 1.9).

Standardized mortality ratio (SMR) data also exist that are consistent with the idea that exposure to low doses of ionizing radiation can be beneficial. The SMR is the ratio of the number of deaths from a specific (defined) cause in one population relative to deaths from the same cause in a second "standard" population, appropriately adjusted for gender, age, and calendar year.[103] Forman et al.[104] reported that SMRs for most cancers were significantly less over a 22 year period in the vicinity of nuclear installations than in non-installation areas. The investigators indicated that this was "unlikely to be due to a protective effect of ionizing radiation" and concluded instead that there were likely to have been large socioeconomic and/or environmental differences between the two groups of SMRs responsible for the apparent anomaly.

A particularly acrimonious controversy centers around cancer SMRs for workers at the Hanford nuclear weapons site in southeastern Washington

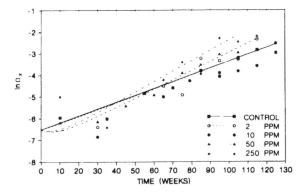

Figure 1.21. Gompertz plots for control and DDT-treated female CF-1 mice. DDT administration (2–250 ppm feed) was begun at 6–7 weeks of age and continued throughout life. Time on the abscissa refers to that period following initiation of dosing. The 2-ppm curve-fit is virtually superimposed on that of controls. Although the data are consistent with concomitant longevity hormesis and toxicity, and this model was used in the curve-fits, the figure is equivocal in making a more definitive judgment. The theoretical lines were obtained by simultaneous fitting of Equations 3 and 12, employing weighted least-squares regression analysis. Both γ and λ were characterized by a logarithmic-logistic function. The original data came from Tomatis et al.[101] Reprinted from Neafsey et al.,[25] p. 143, by permission of Marcel Dekker, Inc.

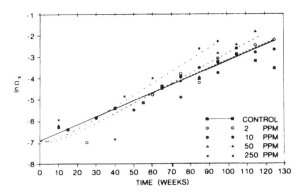

Figure 1.22. Gompertz plots for control and DDT-treated male CF-1 mice. DDT administration (2–250 ppm feed) was begun at 6–7 weeks of age and continued throughout life. Time on the abscissa refers to that period following initiation of dosing. Although the data are consistent with concomitant longevity hormesis and toxicity, and this model was used in the curve-fits, the figure is equivocal in making a more definitive judgment. The theoretical lines were obtained by simultaneous fitting of Equations 3 and 12, employing weighted least-squares regression analysis. Both γ and λ were characterized by a logarithmic-logistic function. The original data came from Tomatis et al.[101] Reprinted from Neafsey et al.,[25] p. 146, by permission of Marcel Dekker, Inc.

state. This site was built in 1943–44 to produce plutonium, which mainly emits γ-radiation. Gilbert et al.[103,105] calculated cancer SMRs for workers at the Hanford site relative to those occurring in the general U.S. population. In general, cancer SMRs were less than unity (averaging 0.85), indicating that Hanford workers were less likely to develop cancer than the U.S. general population. This was attributed to the "healthy worker effect"[106-108] (i.e., employed workers must be able to conduct productive work and are therefore healthier than the general U.S. population). This occurs because the latter population contains individuals unable (too sick) to work and therefore more prone to mortality. Moreover, comparisons of data within the Hanford site showed no evidence of a positive correlation between radiation dose and cancer mortality.

Stewart and Kneale[109] have been highly critical of these findings. They divided the Hanford healthy worker effect into two components:

1. the external healthy worker effect, which asserts that workers were generally healthier than the overall population (see above)
2. the internal healthy worker effect, which asserts that, within this site, healthier workers were more likely to receive jobs exposing them to higher levels of radiation

They point to a strong negative correlation at the Hanford site between radiation dose and overall mortality. In short, they believe that those workers most at risk from cancer due to radiation exposure were unusually healthy (internal and external healthy worker effects), and that this inexorably biases the analysis in favor of reduced mortality among the Hanford workers.

It is not my intention here to sort out all the arguments or to reach any conclusion about the epidemiological issues. Rather, I merely note that a negative correlation between radiation exposure and mortality is consistent with longevity hormesis. If longevity hormesis occurs in human populations exposed to ionizing radiation, the epidemiological consequences are formidable (to say the least!). How does one assimilate (1) an external healthy worker effect, (2) an internal healthy worker effect, (3) toxicity, and (4) longevity hormesis? This conundrum is not exclusive to the radiation industry, but rather permeates most epidemiological studies involving exposures to potentially toxic agents. For example, onsite Australian petroleum industry workers have an SMR (all causes of death) of 0.63 relative to the Australian national population.[110]

CONCLUSION

> There is no love like the old love, that we
> courted in our pride;
> Though our leaves are falling, falling,
> and we're fading side by side,

There are blossoms all around us with the
      colors of our dawn,
And we live in borrowed sunshine when
      the day-star is withdrawn.

Oliver Wendell Holmes, Sr.[111]

If it is assumed that the Napierian logarithm of the hazard function, termed the *Gompertzian,* is proportional to the mean intensity of physiologic injury for homogeneous, laboratory populations of eutherian mammalian species housed under good, uniform laboratory conditions, a simple and convenient method is afforded the investigator to explore the time course of injury accrual and/or disposal. In animal populations, a phenomenon of unknown mechanism, termed *longevity hormesis,* has been demonstrated to reversibly reduce Gompertzians to values below those of control populations. Presumably, this is mediated through a temporarily lessening in one or more types of injury, and is initiated by exposure of organisms to a variety of different stimuli. This response of Gompertzians is distinctly different from that achieved through caloric restriction. Exposures of populations of eutherian mammals to external stimuli, in addition to enhancing life span through longevity hormesis and/or caloric restriction, can also decrease longevity by causing either reversible or irreversible toxicity. Analyses of several data sets suggest that all these actions superimpose their injury promoting or decreasing effects onto the hitherto state of the organism. Evidence for the existence of longevity hormesis in humans is fragmentary and controversial at best.

ACKNOWLEDGMENTS

The author wishes to express his deep appreciation to Frank Hoke, Pat Neafsey, and Gary A. Thompson for their helpful comments in reviewing this chapter.

REFERENCES

1. Larkin, P. *Collected Poems,* Anthony Thwaite, Ed. (London: Farrar, Straus, Giroux, and the Marvell Press, 1989), p. 208.
2. Sacher, G. A. "The Gompertz Transformation in the Study of the Injury-Mortality Relationship: Application to Late Radiation Effects and Ageing," in *Radiation and Ageing,* P. J. Lindop and G. A. Sacher, Eds. (London: Taylor and Francis, 1966), pp. 411–441.
3. Greenwood, M. " 'Laws' of Mortality from the Biological Point of View," *J. Hyg.* 28:267–294 (1928–29).
4. Gompertz, B. "On the Nature of the Function Expressive of the Law of Human Mortality, and on a New Mode of Determining Life Contingencies," *Phil. Trans. Roy. Soc. Lond.* 115A:513–585 (1825); reprinted in part in *Mathe-*

matical Demography, D. Smith and N. Keyfitz, Eds. (Berlin: Springer-Verlag, 1977), Chapter 30, pp. 279-282.

5. Sacher, G. A. "Life Table Modification and Life Prolongation," in *Handbook of the Biology of Aging,* C. E. Finch and L. Hayflick, Eds. (New York: Van Nostrand Reinhold Co., 1977), Chapter 24, pp. 582-638.

6. Sacher, G. A., and R. W. Hart. "Longevity, Aging and Comparative Cellular and Molecular Biology of the House Mouse, *Mus musculus,* and the White-Footed Mouse, *Peromyscus leucopus,*" in *Genetic Effects on Aging,* D. H. Harrison, Ed. (New York: A. R. Liss, 1977), pp. 71-96.

7. Boxenbaum, H. "Gompertz Mortality Analysis: Aging, Longevity Hormesis and Toxicity," *Arch. Gerontol. Geriat.* (in press).

8. Yates, F. E. "The Dynamics of Aging and Time: How Physical Action Implies Social Action," in *Theories of Aging: Psychological and Social Perspectives on Time, Self and Society,* J. E. Birren and V. L. Bengtson, Eds. (New York: Springer, 1988), Chapter 6, pp. 90-117.

9. Witten, M., and C. E. Finch. "Re-Examining the Gompertzian Model of Aging," *IMA Preprint Series #483* (Feb. 1989).

10. Ekonomov, A. L., C. L. Rudd, and A. J. Lomakin. "Actuarial Aging Rate Is Not Constant within the Human Life Span," *Gerontology* 35:113-120 (1989).

11. Jones, H. "Life-Span Studies," in *Basic Mechanisms in Radiobiology. V. Mammalian Aspects,* H. J. Curtis and H. Quastler, Eds. (Washington, DC: National Academy of Sciences–National Research Council, 1957), Chapter IV, pp. 102-170.

12. Strehler, B. L., and A. S. Mildvan. "General Theory of Mortality and Aging," *Science* 132:14-21 (1960).

13. Elandt-Johnson, R. C., and N. L. Johnson. *Survival Models and Data Analysis* (New York: Wiley, 1980).

14. Boxenbaum, H., C. B. McCullough, and F. J. Di Carlo. "Mortality Kinetics in Toxicology Studies," *Drug Met. Rev.* 16:321-362 (1985-86).

15. Sattler, R. *Biophilosophy* (Berlin: Springer-Verlag, 1986), p. 11.

16. Medawar, P. B., and J. S. Medawar. *Aristotle to Zoos* (Cambridge, MA: Harvard University Press, 1983), pp. 5-8, 148-151.

17. Miller, J. G. *Living Systems* (New York: McGraw-Hill Book Co., 1978), p. 90.

18. Boxenbaum, H., P. J. Neafsey, and D. J. Fournier. "Hormesis, Gompertz Functions, and Risk Assessment," *Drug Met. Rev.* 19:195-229 (1988).

19. Wolff, S. "Are Radiation-Induced Effects Hormetic?" *Science* 245:575, 621 (1989).

20. Special Issue on Radiation Hormesis, *Health Phys.* 52(5) (1987).

21. Neafsey, P. J. "Longevity Hormesis. A Review," *Mech. Ageing Dev.* 51:1-31 (1990).

22. Sagan, L. A. "On Radiation, Paradigms, and Hormesis," *Science* 245:574, 621 (1989).

23. Calabrese, E. J., and K. J. Howe. "Stimulation of Growth of Peppermint *(Mentha piperita)* by Phosfon, a Growth Retardant," *Physiol. Plant.* 37:163-165 (1976).

24. Neafsey, P. J., H. Boxenbaum, D. A. Ciraulo, and D. J. Fournier. "A Gompertz Age-Specific Mortality Rate Model of Aging, Hormesis, and Toxicity: Fixed-Dose Studies," *Drug Met. Rev.* 19:369-401 (1988).

25. Neafsey, P. J., H. Boxenbaum, D. A. Ciraulo, and D. J. Fournier. "A Gompertz Age-Specific Mortality Rate Model of Aging, Hormesis, and Toxicity: Dose-Response Studies," *Drug Met. Rev.* 20:111–150 (1989).

26. Aslan, A., A. Vrăbiescu, C. Domilescu, L. Cámpeanu, M. Costiniu, and S. Stănescu. "Long-Term Treatment with Procaine (Gerovital H₃) in Albino Rats," *J. Gerontol.* 20:1–8 (1965).

27. Neafsey, P. J. "Age-Specific Mortality Rate Analysis: A New Paradigm for Risk Assessment," *Drug Met. Rev.* 21:471–485 (1989–90).

28. Neafsey, P. J., H. Boxenbaum, D. A. Ciraulo, and D. J. Fournier. "A Gompertz Age-Specific Mortality Rate Model of Aging: Modification by Dietary Restriction in Rats," *Drug Met. Rev.* 21:351–365 (1989).

29. *Rubáiyát of Omar Khayyám,* rendered into English verse by E. Fitzgerald (Garden City, NY: Doubleday and Co., 1952).

30. Yates, F. E. "Evolutionary Computing by Dynamics in Living Organisms," in *Advances in Cognitive Science: Steps toward Convergence,* M. Kochen and H. M. Hastings, Eds. (Boulder, CO: Westview Press, 1988), Chapter 3, pp. 26–49.

31. Asimov, I., and J. A. Shulman, Eds. *Isaac Asimov's Book of Science and Nature Quotations* (New York: Weidenfeld and Nicolson, 1988), p. 34.

32. Cutler, R. G. "Antioxidants, Aging, and Longevity," in *Free Radicals in Biology,* Vol. VI, W. A. Pryor, Ed. (New York: Academic Press, 1984), Chapter 11, pp. 371–428.

33. Bass, L., H. S. Green, and H. Boxenbaum. "Gompertzian Mortality Derived from Competition between Cell-Types: Congenital, Toxicologic and Biometric Determinants of Longevity," *J. Theor. Biol.* 140:263–278 (1989).

34. Meites, J. "Aging Studies," *Science* 251:855 (1991).

35. Sacher, G. A. "Problems in the Extrapolation of Long-Term Effects from Animals to Man," in *The Delayed Effects of Whole-Body Radiation,* B. B. Watson, Ed. (Baltimore: The Johns Hopkins Press, 1960), pp. 3–10.

36. Sacher, G. A. "On the Statistical Nature of Mortality, with Especial Reference to Chronic Radiation Mortality," *Radiology* 67:250–258 (1956).

37. Sacher, G. A. "Lethal Effects of Whole-Body Irradiation in Mice: The Dose-Time Relation for Terminated and Time-Dependent Exposure," *J. Nat. Canc. Inst.* 32:227–241 (1964).

38. Sacher, G. A. "Dose, Dose Rate, Radiation Quality, and Host Factors for Radiation-Induced Life Shortening," in *Aging, Carcinogenesis, and Radiation Biology,* K. S. Smith, Ed. (New York: Plenum, 1976), pp. 493–517.

39. Sacher, G. A., and E. Trucco. "The Stochastic Theory of Mortality," *Ann. N.Y. Acad. Sci.* 96:985–1007 (1962).

40. Lansdown, A. B. G., and D. M. Conning. "Animal Husbandry," in *Experimental Toxicology,* D. Anderson and D. M. Conning, Eds. (London: Royal Society of Chemistry, 1988), Chapter 6, pp. 82–106.

41. Lee, E. T. *Statistical Methods for Survival Data Analysis* (Belmont, CA: Lifetime Learning Publications, 1980).

42. Weibull, W. "A Statistical Distribution Function of Wide Applicability," *J. Appl. Mech.* 18:293–297 (1951).

43. Sacher, G. A. "Reparable and Irreparable Injury: A Survey of the Position in Experiment and Theory," in *Radiation Biology and Medicine,* W. D. Claus, Ed. (Reading, MA: Addison-Wesley, 1958), Chapter 12, pp. 283–313.

44. Sacher, G. A. "Models from Radiation Toxicity Data," in *Late Effects of Radiation*, R. J. M. Fry, D. Grahn, M. L. Griem, and J. H. Rust, Eds. (London: Taylor and Francis, 1970), Chapter X, pp. 233–244.

45. Thron, C. D. "Linearity and Superposition in Pharmacokinetics," *Pharmacol. Rev.* 26:3–31 (1974).

46. Boxenbaum, H., and D. Ciraulo. "Alcoholism and Mortality Kinetics," *J. Pharm. Sci.* 73:1191–1195 (1984).

47. Thompson, G. A., J. Smithers, and H. Boxenbaum. "Biphasic Mortality Response of Chipmunks in the Wild to Single Doses of Ionizing Radiation: Toxicity and Longevity Hormesis," *Drug Met. Rev.* 22:269–289 (1990).

48. Anderson, K. M., W. P. Castelli, and D. Levy. "Cholesterol and Mortality: 30 Years of Follow-Up from the Framingham Study," *J. Am. Med. Assoc.* 257:2176–2180 (1987).

49. Gibbons, A. "Gerontology Research Comes of Age," *Science* 250:622–625 (1990).

50. Boxenbaum, H., and R. W. D'Souza. "Interspecies Pharmacokinetic Scaling, Biological Design and Neoteny," in *Advances in Drug Research,* Vol. 19, B. Testa, Ed. (London: Academic Press, 1990), pp. 139–196.

51. Calder, W.A., III. *Size, Function, and Life History* (Cambridge, MA: Harvard University Press, 1984).

52. Gould, S. J. *Ontogeny and Phylogeny* (Cambridge, MA: The Belknap Press of Harvard University Press, 1977).

53. Schmidt-Nielsen, K. *Scaling: Why Is Animal Size So Important?* (Cambridge: Cambridge University Press, 1984).

54. Snyder, W. S., M. J. Cook, E. S. Nasset, L. R. Karhausen, G. P. Howells, and I. H. Tipton (Commissioners). *Report of the Task Group on Reference Man* (New York: Pergamon Press, 1975), p. 213.

55. Montagu, A. *Growing Young,* 2nd ed. (New York: Bergin and Garvey Publishers, 1989).

56. Gould, S. J. *The Panda's Thumb* (New York: W.W. Norton and Co., 1980).

57. Naef, A. "Über die Urformen der Anthropomorphen und die Stammesgeschichte des Menschenschädels," *Naturwissenschaften* 14:445–452 (1926).

58. Lewontin, R. C. "Adaptation," *Sci. Am.* 239(3):212–230 (1978).

59. Brown, N. O. *Life Against Death* (Middletown, CT: Wesleyan University Press, 1959), pp. 25–31.

60. Haldane, J. B. S. "Time Is Biology," *Science Prog.* 44(July):385–402 (1956).

61. Toulmin, S. "The Construal of Reality: Criticism in Modern and Postmodern Science," in *The Politics of Interpretation,* W. J. T. Mitchell, Ed. (Chicago: The University of Chicago Press, 1983), pp. 99–117.

62. Waddington, C. H. *The Nature of Life* (London: George Allen and Unwin, 1961).

63. Pool, R. "Science's Top 20 Greatest Hits," *Science* 251:267 (1991); citing unpublished views of R. Hazen and J. Trefil.

64. Gibaldi, M., and D. Perrier. *Pharmacokinetics,* 2nd ed. (New York: Marcel Dekker, 1982), Chapter 6, pp. 221–269.

65. Holford, N. H. G., and L. B. Sheiner. "Understanding the Dose-Effect Relationship: Clinical Application of Pharmacokinetic-Pharmacodynamic Models," *Clin. Pharmacokin.* 6:429–453 (1981).

66. Holford, N. H. G., and L. B. Sheiner. "Pharmacokinetic and Pharmacodynamic Modeling in vivo," *CRC Crit. Rev. Bioeng.* 5:273–322 (1981).

67. Holford, N. H. G., and L. B. Sheiner. "Kinetics of Pharmacologic Response," *Pharmacol. Ther.* 16:143–166 (1982).

68. Kenakin, T. P. *Pharmacologic Analysis of Drug-Receptor Interaction* (New York: Raven Press, 1987).

69. Levitzki, A. *Receptors: A Quantitative Approach* (Menlo Park, CA: The Benjamin/Cummings Pub. Co., 1984).

70. Riggs, D. S. *The Mathematical Approach to Physiological Problems* (Baltimore: The Williams and Wilkins Co., 1963), Chapter 11, pp. 272–298.

71. Tallarida, R. J., and L. S. Jacob. *The Dose-Response Relation in Pharmacology* (New York: Springer-Verlag, 1979).

72. Wagner, J. G. "Kinetics of Pharmacologic Response. I. Proposed Relationships between Response and Drug Concentration in the Intact Animal and Man," *J. Theor. Biol.* 20:173–201 (1968).

73. Wagner, J. G. "Aspects of Pharmacokinetics and Biopharmaceutics in Relation to Drug Activity," *Am. J. Pharm.* 141:5–20 (1969).

74. Wagner, J. G. "Relations between Drug Concentration and Response," *J. Mond. Pharm.* 4:279–310 (1971).

75. Boxenbaum, H. G., S. Riegelman, and R. M. Elashoff. "Statistical Estimations in Pharmacokinetics," *J. Pharmacokin. Biopharm.* 2:123–148 (1974).

76. Heidrick, M. L., L. C. Hendricks, and D. E. Cook. "Effect of Dietary 2-Mercaptoethanol on the Life Span, Immune System, Tumor Incidence and Lipid Peroxidation Damage in Spleen Lymphocytes of Aging BC3F1 Mice," *Mech. Ageing Dev.* 27:341–358 (1984).

77. Skalicky, M., H. Bubna-Littitz, and G. Hofecker. "The Influence of Persistent Crowding on the Age Changes of Behavioral Parameters and Survival Characteristics of Rats," *Mech. Aging Dev.* 28:325–336 (1984).

78. Sacher, G. A. "Effects of X-Rays on the Survival of *Drosophila* Imagoes," *Physiol. Zool.* 36:295–311 (1963).

79. Paracelsus. "Seven Defensiones: The Reply to Certain Calumniations of His Enemies," C. L. Temkin (translator), in *Four Treatises of Theophrastus von Hohenheim called Paracelsus,* H. E. Sigerist, Ed. (Baltimore: The Johns Hopkins Press, 1941), pp. 1–41.

80. Daniel, C., and F. S. Wood. *Fitting Equations to Data,* 2nd ed. (New York: John Wiley and Sons, 1980).

81. Brownlee, K. A. *Statistical Theory and Methodology in Science and Engineering* (New York: Wiley, 1960), pp. 306–307.

82. Yu, B. P., E. J. Masoro, and C. A. McMahan. "Nutritional Influences on Aging of Fischer 344 Rats: I. Physical, Metabolic, and Longevity Characteristics," *J. Gerontol.* 40:657–670 (1985).

83. Lorenz, E., J. W. Hollcroft, E. Miller, C. C. Congdon, and R. Schweisthal. "Long-Term Effects of Acute and Chronic Irradiation in Mice. I. Survival and Tumor Incidence Following Chronic Irradiation of 0.11 r per Day," *J. Nat. Canc. Inst.* 15:1049–1058 (1955).

84. Lorenz, E., L. O. Jacobson, W. E. Heston, M. Shimkin, A. B. Eschenbrenner, M. K. Deringer, J. Doniger, and R. Schweisthal. "Effects of Long-Continued Total-Body Gamma Irradiation on Mice, Guinea Pigs, and Rabbits. III. Effects of Life Span, Weight, Blood Picture, and Carcinogenesis and the Role

of the Intensity of Radiation," in *Biological Effects of External X and Gamma Radiation,* Part 1, R. E. Zirkle, Ed. (New York: McGraw Hill Book Co., 1954), Chapter 3, pp. 24–148.

85. Smyth, H.F, Jr. "Sufficient Challenge," *Food Cosmet. Toxicol.* 5:51–58 (1967).

86. Ottoboni, M. A. *The Dose Makes the Poison* (Berkeley, CA: Vincente Books, 1984), pp. 91–95.

87. Newton-Smith, W. H. *The Rationality of Science* (Boston: Routledge and Kegan Paul, 1981).

88. Star, S. L. "Scientific Work and Uncertainty," *Soc. Stud. Sci.* 15:391–427 (1985).

89. Becker, E. *The Structure of Evil* (New York: The Free Press, 1968).

90. Eckberg, D. L., and L. Hill, Jr. "The Paradigm Concept and Sociology: A Critical Review," in *Paradigms and Revolutions,* G. Gutting, Ed. (Notre Dame, IN: University of Notre Dame Press, 1980), pp. 117–136.

91. Cember, H. *Introduction to Health Physics,* 2nd ed. (New York: Pergamon Press, 1983), p. 142.

92. Tryon, C. A., and D. P. Snyder. "The Effect of Exposure to 200 and 400 R of Ionizing Radiation on the Survivorship Curves of the Eastern Chipmunk *(Tamias striatus)* under Natural Conditions," in *Radionuclides in Ecosystems,* D. J. Nelson, Ed. (Springfield, VA: National Technical Information Service, 1973), pp. 1037–1041.

93. Dunaway, P. B., L. L. Lewis, J. D. Story, J. A. Payne, and J. M. Inglis. "Radiation Effects in the *Soricidae, Cricetidae,* and *Muridae,*" in *Symposium on Radioecology,* D. J. Nelson and F. C. Evans, Eds. (Washington, DC: Fed. Sci. and Tech. Info., U.S.D.C., 1969), pp. 173–184.

94. Burek, J. D., K. D. Nitschke, T. J. Bell, D. L. Wackerle, R. C. Childs, J. E. Beyer, D. A. Dittenber, L. W. Rampy, and M. J. McKenna. "Methylene Chloride: A Two-Year Inhalation Toxicity and Oncogenicity Study in Rats and Hamsters," *Fund. Appl. Toxicol.* 4:30–47 (1984).

95. "Carcinogenesis Bioassay of Amosite Asbestos (CAS No. 12172–73–5) in Fischer 344/N rats (Feed Study)," National Toxicology Program Technical Report No. 279 (Draft), U.S. Department of Health and Human Services, Washington, DC (1984).

96. "Lifetime Carcinogenesis Studies of Amosite Asbestos (CAS No. 121–72–73–5) in Syrian Golden Hamsters (Feed Studies)," National Toxicology Program Technical Report No. 249, U.S. Department of Health and Human Services, Washington, DC (1983).

97. Walker, A. I. T., E. Thorpe, and D. E. Stevenson. "The Toxicology of Dieldrin (HEOD). I. Long-Term Oral Toxity Studies in Mice," *Food Cosmet. Toxicol.* 11:415–432 (1973).

98. Miller, R. R., J. T. Young, R. J. Kociba, D. G. Keyes, K. M. Bodner, L. L. Calhoun, and J. A. Ayres. "Chronic Toxicity and Oncogenicity Bioassay of Inhaled Ethyl Acrylate in Fischer 344 Rats and B6C3F1 Mice," *Drug Chem. Toxicol.* 8:1–42 (1985).

99. Jorgenson, T. A., E. F. Meierhenry, C. J. Rushbrook, R. J. Bull, and M. Robinson. "Carcinogenicity of Chloroform in Drinking Water to Male Osborne-Mendel Rats and Female B6C3F1 Mice," *Fund. Appl. Toxicol.* 5:760–769 (1985).

100. Arnold, D. L., C. A. Moodie, S. M. Charbonneau, H. C. Grice, P. F. McGuire, F. R. Bryce, B. T. Collins, Z. Z. Zawidzka, D. R. Krewski, E. A. Nera, and I. C. Munro. "Long-Term Toxicity of Hexachlorobenzene in the Rat and the Effect of Dietary Vitamin A," *Food Chem. Toxicol.* 23:779–793 (1985).

101. Tomatis, L., V. Turusov, N. Day, and R. T. Charles. "The Effect of Long-Term Exposure to DDT on CF-1 Mice," *Int. J. Canc.* 10:489–506 (1972).

102. Matanoski, G. M., A. Sternberg, and E. A. Elliott. "Does Radiation Exposure Produce a Protective Effect among Radiologists?" *Health Phys.* 52:637–643 (1987).

103. Gilbert, E. S., G. R. Petersen, and J. A. Buchanan. "Mortality of Workers at the Hanford Site: 1945–1981," *Health Phys.* 56:11–25 (1989).

104. Forman, D., P. Cook-Mozaffari, S. Darby, G. Davey, I. Stratton, R. Doll, and M. Pike. "Cancer near Nuclear Installations," *Nature* 329:499–505 (1987).

105. Gilbert, E. S., G. R. Petersen, and J. A. Buchanan. "Mortality of Hanford Workers: A Reply," *Health Phys.* 57:841 (1989).

106. McMichael, A. J. "Standardized Mortality Ratios and the 'Healthy Worker Effect': Scratching beneath the Surface," *J. Occup. Med.* 18:165–168 (1976).

107. McMichael, A. J., R. Spirtas, and L. L. Kupper. "An Epidemiologic Study of Mortality within a Cohort of Rubber Workers, 1964–72," *J. Occup. Med.* 16:458–464 (1974).

108. Weed, D. L. "Historical Roots of the Healthy Worker Effect," *J. Occup. Med.* 28:343–347 (1986).

109. Stewart, A., and G. Kneale. "Mortality of Hanford Workers," *Health Phys.* 57:839–841 (1989).

110. Christie, D., K. Robinson, I. Gordon, C. Webley, and J. Bisby. "Current Mortality in the Australian Petroleum Industry: The Healthy-Worker Effect and the Influence of Life-Style Factors," *Med. J. Aust.* 147:222–225 (1987).

111. Holmes, O. W. "*The Poetical Works of Oliver Wendell Holmes,* (Boston: Houghton, Mifflin, and Co., 1890), p. 298.

Note Added in Proof: Subsequent to submitting this chapter, I became aware of two papers which seriously challenge the claim that neoteny has played a key role in human evolution [Shea, B.T. "Heterochrony in Human Evolution: The Case for Neoteny Reconsidered," *Yearbook of Physical Anthropology* 32:69–101 (1989) and Deacon, T.W. "Problems of Ontogeny and Phylogeny in Brainsize Evolution," *Int. J. Primatol.* 11:237–282 (1990)]. It thus appears that other evolutionary (phylogenetic and ontogenetic) mechanisms are primarily responsible for the unusually slow rate of human aging discussed in the context of Hypothesis VIII.

CHAPTER 2

The Role of the "Stress Protein Response" in Hormesis

Joan Smith-Sonneborn, Zoology and Physiology Department, University of Wyoming, Laramie, Wyoming

Hormesis refers to the phenomenon of induction of beneficial effects by low doses of otherwise harmful physical or chemical agents:[1] "a little bit of bad can be good for you." That the hormetic response may operate by a common mechanism already has been proposed,[2,3] but this review is the first to propose the hypothesis that the common pathway is a heat shock–like response. The heat shock response is a model for a more general phenomenon, called "the stress response." The stress response is characterized by increased synthesis of a family of stressor specific proteins with concomitant reduction of synthesis of most of the proteins transcribed prior to the exposure to the toxic agent.[4] The stress response has been characterized using heat, radiation, heavy metals, and oxidizing agents as the stressors.[5]

To develop the hypothesis that the hormetic response may operate through the stress response, this chapter includes

1. identification of agents known to induce both the stress response and hormetic phenomena
2. a description of the unique and common pathways in the stress response to three stressors: heat, DNA-damaging agents, and teratogens
3. the stress response as a model for teratogen-induced damage
4. a theory explaining the paradoxical beneficial response to low doses of an otherwise harmful agent via a stress-response pathway

HORMETIC AGENTS AND STRESS INDUCERS

Hormetic agents are highly diverse, including heavy metals, polychlorinated biphenyls, insecticides, alcohol, oxygen poisoning, cyanide, antibiotics,[6] ionizing radiation,[7] cosmic or gamma radiation,[8,9] electromagnetic radiations,[10,11] and ultraviolet plus photoreactivation.[12] Examples of beneficial

Table 2.1. Agents Identified As Both Hormetic and Heat Shock Agents

Hormetic and Heat Shock Agents	Beneficial Response	References
Cadmium	increased survival	6,13
	increased growth	6
	increased hormone secretion	6
	increased acclimation	14
Mercury	increased survival	6,13
	increased growth	6
	increased ATPase	6
Copper	increased survival	6,13
	increased growth	6,13
Zinc	increased survival	6,13
	increased growth	6,13
Ethanol	behavior	6,13
Oxygen poison	vital signs improved	6,13
Chloramphenicol	increased growth	6,13
X-rays	Increased mean life span	7,15
	faster seed growth	16
Ultraviolet radiation	life span (UV + PR)	12,17,18
Heat	suboptimal benefit	19,20,21

biological responses include increased life span, cell division rate,[10,12] accelerated maturation time, and acclimation (see Table 2.1).[6,7,12-21]

Agents identified as both hormetic agents and inducers of the stress response are listed in Table 2.1. There is an impressive overlap between the hormetic and stress response inducers, though no experiment was designed to correlate induction of the stress response and the onset of a beneficial biological effect. However, a biomarker of the stress response is induced resistance to the stressor. The induction of resistance or acclimation to challenge with higher doses or prolonged exposure to the stressing agents include the following:

1. Prior heat treatment can induce survival of organisms to higher temperatures.[20,22]
2. Prior heat treatment can prevent heat-induced developmental defects.[21]
3. Hormetic agents like cadmium and ethanol can induce cross-resistance to other environmental stressors like heat-inducing thermotolerance.[20]
4. In *E. coli* a chemical mutagen, MNNG, can induce resistance after the first hour of treatment by induction of a novel form of repair.[23]
5. A more youthful resistance to ultraviolet irradiation was found in *Parame-*

cium exposed to a prior regime of ultraviolet and photoreactivation treatment at doses that affected an increased mean life span.[12]

6. Fish with prior exposure to cadmium could regenerate clipped fins faster than nonexposed organisms in cadmium-contaminated water.[14]

Since not all stressors induce the same transcripts or metabolic changes (see below), all stressors need not be hormetic agents. However, this proposal predicts that seemingly unrelated agents that induce the same stress response should stimulate the same biological effects and induce cross-resistance. Likewise, stressors with opposing alterations in chromatin should increase sensitivity, not resistance, to the stressing agent.

Available data on the molecular biology of the stress response to various stressors is reviewed below as a potential model of a hormetic pathway.

THE STRESS RESPONSE

The stress proteins are divided into two groups: those referred to as the heat shock proteins, first found induced by nonphysiological exposure to heat; and those called the glucose-regulated proteins, which exhibit increased synthesis when cells are deprived of glucose or oxygen, or when calcium homeostasis is disrupted. Members of the two families exhibit considerable homology.[4]

Heat

The first stress response detected was appearance of puffs induced in the salivary gland of fruit flies by heat and dinitrophenol.[24] The heat shock response is universal from bacteria to humans.[13,20,25] The stress response induced by heat is characterized by transcription of a coordinately regulated subset of induced proteins, repression of the transcription, and translation of previously active genes and preexisting messages.[13] The heat shock proteins are members of families of proteins with species-related molecular weight range classes. In higher organisms, the high molecular weight families are Hsp 110, a normal nucleolar protein found in vertebrates; the Hsp 100 family (Mr 92–102), a phosphoprotein normally present in the plasma membrane; Hsp 89 (Mr 83–95), found in the soluble protein in all animal cells; and the Hsp 70 family (Mr 68–78).[4,25] The multigene Hsp 70 family has members in the cytoplasm, in the lumen of the endoplasmic reticulum, and in the matrix of mitochondria which function in protein translocation across membranes.[26]

The heat shock protein Hsp 70 is a major factor in the heat response since

1. mammalian cells in which Hsp 70 is not made or is inactivated by antibody binding cannot develop thermotolerance[27]

2. cycloheximide can induce tolerance to higher temperatures, but heat shock proteins are required for full protection[28]
3. the thermolability of mouse oocytes is due to the lack of expression and/or inducibility of Hsp 70[29]

Smaller Hsp's (15–28 kd) bind reversibly to the nuclear skeleton during heat shock and form higher-order aggregates. A common central domain of the four small *Drosophila* Hsp's[22,23,26,28] show great similarity to alpha crystallin.[25,30]

The smallest Hsp's are the ubiquitin family (7–8 kd),[31] which have been implicated as regulator molecules in chromatin, DNA repair, meiosis, sporulation, degradation of abnormal proteins,[32] ribosome biogenesis,[33] and facilitation of transposition.[34]

Besides the induction of heat shock proteins, other metabolic changes found in response to nonphysiological heat exposure, which impact on chromatin structure, include

1. increased levels of high molecular weight ubiquitin conjugates and decreased ubiquitinated histone in HeLa cells[35,36] (ubiquinated DNA is associated with active expression)[37]
2. hypermethylation of H2 B and decreased methylation of H3[38]
3. the ubiquinated form of histones in yeast when grown under mildly stressful but not lethal temperature[35]

Topological changes in chromatin are typical of the heat shock response[39] and are assumed to participate in the changes in heat-induced gene expression and repression.

A presumed physiological consequence of heat-induced alteration of chromatin is heat-induced radiosensitivity of cancer cells.[40,41] Heat induces a dramatic increase of nonhistone protein content, resulting in a reduced affinity to repair enzymes.[40]

Heat also causes conformational changes of membrane lipids and proteins,[42,43] excessive fluidization of the plasma membrane, and leakage of required low molecular weight components.[43] Low doses of local anesthetics procaine and lidocaine, known to decrease membrane viscosity, increase neoplastic killing.[43] The membrane defects may cause release of polyamines and disturb DNA replication.[40,41]

DNA-Damaging Agents

In prokaryotes' response to a given stressor, unlinked and individually controlled genes can be coordinately controlled by common regulator genes called *regulons*.[5] The damage response in bacteria to ultraviolet irradiation is the "SOS" response;[44,45] to reactive oxygen species, the oxy R response;[46] and specialized responses to other environmental stresses, like cold, heat, nutrient limitation, salinity, and osmolarity, are well characterized.[47,48] Different stressors are related in the sense that they share member genes or

Table 2.2. DNA Damage Response

Induced Proteins

Plasminogen activator, a protease[49]
PolyADP ribose[50]
DNA ligase[51,52]
Metallothione[53–55]
H2 antigen[56]
Extracellular inducing factor[57,58]
Collogenase[53]
c fos[53]
c myc[59]
p53 tumor antigen[60]
DNA polymerase[61,62]
Hsp 28[18]

Metabolic Changes

Inhibition of DNA methylation[63]
Demethylation[55]

protein products that interact. For example, in *Escherichia coli*, both heat and ethanol initiate the same response (i.e., solely a heat shock response). On the other hand, both hydrogen peroxide and 6 amino-7-chloro-5,8-dioxoquinoline (ACDQ) stimulate an oxidation stress response and a secondary SOS response; nalidixic acid and puromycin, an SOS and heat shock response; isoleucine restriction, a poor heat shock response; and cadmium chloride strongly induces all three stress responses.[5] The regulon typical response to ACDQ, cadmium chloride, and hydrogen peroxide was a minor response; these agents stimulated the synthesis of another 35 proteins by 5- to 50-fold. Another accumulated product of exposure to certain stressors are adenylated nucleotides, which are candidates as alarmones.[5]

Thus, general and specific cellular responses are triggered by different stressors. Ultraviolet- or carcinogen-related DNA damage-induced expression of the stress response does not appear to conform to the prokaryote SOS model.[48] DNA-damaging agents induce a spectrum of molecular responses, including the production of proteases, DNA repair agents, oncogenes, and chromatin changes (Table 2.2).[18,49-63] One gene product induced, extracellular inducing factor (EPIF), can induce the ultraviolet spectrum of proteins in untreated cells.[57,58] The induction of a hormetic effect by EPIF would shed light on the participation of these gene products in the protective pathway. Besides the induction of specific identified (and unidentified) proteins, a major change induced by DNA damage is alteration of the chromatin structure involving increased synthesis of poly (ADP-ribose),[50] alteration of histone methylation patterns,[55,63] and dependence on the presence of ubiquitin-histone conjugants.[31] In contrast with heat shock, DNA-damaging agents inhibit rather than increase DNA methylation,[63] and/or induce demethylation.[55] The ubiquitin conjugating enzyme is essential for DNA repair since loss of the ubiquitin-conjugating enzyme, E3, results in slow growth; sensitivity to UV, X-rays, and chemical mutagens; retrotransposition; and inability to sporulate.[31] A suggested role for the ubiquitin-conjugating enzyme is to mediate changes in chromatin by patched degradation of chromosomal proteins to allow access for repair.[31]

Degradation of Abnormal Proteins Produced by Stressors

When cells are exposed to heat and other toxic agents, abnormal proteins accumulate. The abnormal proteins signal expression of heat shock proteins, which can directly interact with the protein for "protein repair" by catalyzing ATP-driven refolding.[64] The unrepaired proteins are eliminated by a second major pathway of the response, an ATP-driven elimination of abnormal proteins mediated by the ubiquitin system. But imbalances in the protein degradation system, perhaps induced by an overload of abnormal proteins, can result in premature degradation of necessary regulatory molecules.[65] With respect to hormesis, the beneficial stress response may be protein repair and the elimination of abnormal proteins. The detrimental response may be the inappropriate destruction of short-lived essential regulator molecules when a threshold level of abnormal proteins are produced by toxic agents, radiation damage, aging, or age-related diseases. In addition to imbalance in the degradation pathway at higher doses when abnormal proteins accumulate, changes in the fundamental structure of the essential ubiquitin-conjugating enzymes is dosage dependent, at least with respect to heat.[31] The ubiquitin-conjugating enzymes, essential for survival to the stressing agent, have introns. Since splicing of introns is blocked at higher temperatures,[66,67] the introns could serve to restrict function of protective enzymes to moderate, not severe stress.[31]

Heat Shock Genes in the DNA Damage Response

The role of heat shock genes in the DNA damage response is not known. But heat shock genes do appear in the DNA damage response. Hsp 70 expression is temporarily correlated with maximal survival of viruses after UV irradiation of viral infected cells,[15] and small Hsp's were induced by UV and teratogenic agents.[15,68] Ubiquitin was induced after treatment with mutagens and teratogens.[69-71]

Using ionizing irradiation of rat embryos in utero, enhanced expression of Hsp 70 and c-myc was increased 4 or 5 days after treatment, and c-fos increased only after the embryos were incubated in vitro.[15,72] Coordinate expression of Hsp 70 and c-myc have been detected during heat shock.[73]

Chemical teratogens showed enhanced induction of small heat shock proteins in embryos when cultivated in vitro[72] and induced a subset of small heat shock protein in flies,[71,74] and ubiquitin in mammalian cells.[75,76]

Since the ubiquitinated Rad6 DNA repair enzyme is a ubiquitin-conjugating enzyme essential for normal growth, sporulation, and repair,[31] ubiquitin may be a key regulatory molecule in the stress response. Changes in metabolism of ubiquitin, as well as increased synthesis of unique forms of ubiquitin gene family members, may shed light on controlling elements.

Teratogenic Agents and Heat Shock Agents

The common pathway of several apparently unrelated chemicals (and heat) to induce the teratogenic response was reviewed.[21] The known teratogenic agents — heat, ethanol, arsenite, cortisone, retinoic acid, valproic acid, cadmium, diazepam, verapamil, and phenobarbital — all induce some or all members of the so-called heat shock proteins.[21] Cadmium and ethanol are also hormetic agents. The type of defect induced during development by the teratogenic agent depends critically on the timing of the environmental insult.[77] Defects induced by heat in flies can be induced only at specific times during development. Heat treatment can alter the order of development time. During recovery, heat shock proteins are synthesized first, then synthesis and decay of messages involved in the developmental program.[78] The interruption in the development and delay in resumption can cause the failure to complete one process before the next process begins.[79] In mammals, teratogens induce heat shock protein and affect differentiation of nerve and muscle in *Drosophila* embryonic cells.[78,79]

Molecular Models of Developmental Defects

The common pathway then is the interruption of an ordered series of events by any chemical or physical agent that induces a stress response. The stress response is a cessation of the synthesis of normal proteins with the selective production of the proteins required to cope with the specific toxin. The interruption, not the agent, triggers the defect. The timing of the insult dictates the defect. Recovery depends on the length and severity of treatment as well as on whether the temperature is raised slowly or abruptly.[22]

Stress Proteins: Resolution of the Paradox

The hypothesis that the stress protein response is the common pathway for hormetic agents is supported by the following findings:

1. The same agents identified as hormetic also induce the stress response.
2. Some hormetic agents with molecular responses common to the heat shock response can induce thermotolerance, while others with known differences in induction of methylation patterns induce sensitivity.
3. The stress response includes preferential synthesis of products that repair both protein and DNA, which could stimulate growth and longevity.
4. The alterations in the chromatin structure could facilitate derepression of growth-promoting products or provide access to DNA for repair.

There is a model for a biphasic response using heat as the stressor. In moderate doses, the protective molecular reactions progress. But at higher temperatures, intron splicing is inhibited, and therefore production of the needed protective response. Other as yet unknown important differences in molecular responses at low and high doses may be uncovered in the future.

In summary, the stress response could provide an explanation for a beneficial response to an otherwise harmful agent. The potential for a theoretical biological beneficial response stems from the induction of cellular repair processes. The protective responses include

1. expression of "protein repair" proteins, like the heat shock proteins, which can monitor proper folding of denatured proteins
2. stimulation of elimination of abnormal proteins that cannot be repaired
3. induction of increased DNA repair and replication molecules
4. alteration of chromatin structure to facilitate repair of regions previously refractory to repair and/or alter gene expression to accelerate growth and maturation
5. induction of cross-resistance to other environmental toxins, thereby increasing tolerance to the same or apparently unrelated environmental toxins that are life-shortening agents

Why the beneficial response is effected only at low doses cannot yet be explained, but the inability to remove introns from gene transcripts required for survival, at moderate but not high temperatures, and changes in histone-ubiquitin conjugates may provide a clue to explain cytotoxic and genotoxic responses after a threshold limit for a beneficial response.

Since different stressors have specific responses, not all stressors are expected to be beneficial — or beneficial with respect to the same parameter. The hormetic response may not be an "overcorrection" response to the damaging agent, but rather a benefit derived from the "stress" response (i.e., repair or removal of accumulated age or environmental induced cellular damage in proteins, genes, and cell membranes; chromatin changes to accelerate seed maturation; or cross-resistance to certain other environmental toxins).

REFERENCES

1. *Health Phys.* 52:517–680 (1987).
2. Luckey, T. D. "Ionizing Radiation Promotes Protozoan Reproduction," *Radiat. Res.* 108:215–221 (1986).
3. Stebbing, A. R. D. "Growth Hormesis: A Byproduct of Control," *Health Phys.* 52:543–548 (1987).
4. Welch, W. J., L. A. Mizzen, and A. P. Arrigo. "Structure and Function of Mammalian Stress Proteins," in *Proteins,* M. L. Pardue, J. R. Feramisco, and S. Lindquist, Eds. (New York: Alan R. Liss, 1989), p. 187.
5. VanBogelen, R., P. M. Kelley, and F. C. Neidhardt. "Differential Induction of Heat Shock, SOS and Oxidation Stress Regulons and Accumulation of Nucleotides in *Escherichia coli,*" *Bacteriol.* 169:26–32 (1987).
6. Calabrese, E. J., M. E. McCarthy, and E. Kenyon. "The Occurrence of Chemically Induced Hormesis," *Health Phys.* 52:531–542 (1987).

7. Congdon, C. C. "A Review of Certain Low-Level Ionizing Radiation Studies in Mice and Guinea Pigs," *Health Phys.* 52:93–598 (1987).

8. Planel, H., R. Soleilhavoup, and R. Tixador. "Influence of Cell Proliferation on Background Radiation or Exposure to Very Low Chronic Gamma Radiation," *Health Phys.* 52:571–578 (1987).

9. Tixador, R., G. Richoiley, E. Monrozies, H. Planel, and G. Tap. "Effects of Very Low Doses of Ionizing Radiation on the Clonal Life-Span in *Paramecium tetraurelia,*" *Int. J. Radiat. Biol.* 39:47–54 (1981).

10. Darnell, C. "Effects of Extremely Low Electromagnetic Radiation on Paramecium Life Span and Ion Conductance," MS Thesis, University of Wyoming, Laramie, WY (1988).

11. Dihel, L., and J. Smith-Sonneborn. "Effects of Low Frequency Electromagnetic Field on Cell Division and the Plasma Membrane," *Bioelectromagnetics* 6:61–71 (1985).

12. Smith-Sonneborn, J. "DNA Repair and Longevity Assurance in *Paramecium tetraurelia,*" *Science* 203:1115–1117 (1979).

13. Nover, L. *Heat Shock: Response of Eukaryotic Cells* (New York: Springer-Verlag, 1984), pp. 1–82.

14. Weis, P., and J. S. Weis. "Cadmium Acclimation and Hormesis in *Fundulus heteroclitus* During Fin Regeneration," *Environ. Res.* 39:356–363 (1986).

15. Higo, H., J. Y. Lee, Y. Satow, and K. Higo. "Elevated Expression of Proto-Oncogenes Accompany Enhanced Induction of Heat-Shock Genes after Exposure of Rat Embryos In Utero to Ionizing Irradiation," *Teratogenesis, Carcinogenesis, and Mutagenesis* 9:191–198 (1989).

16. Sheppard, S. C., and P. J. Regitnig. "Factors Controlling the Hormesis Response in Irradiated Seed," *Health Phys.* 52:599–606 (1987).

17. Williams, K. J., B. Landgraf, N. L. Whiting, and J. Zurlo. "Correlation Between the Induction of Heat Shock Protein 70 and Enhanced Viral Reactivation in Mammalian Cells Treated with Ultraviolet Light and Heat Shock," *Cancer Res.* 49:2735–2742 (1989).

18. Vivino, A. A., M. D. Smith, and K. W. Minton. "A DNA Damage-Responsive *Drosophila melanogaster* Gene Is Also Induced by Heat Shock," *Mol. Cell Biol.* 6:4767–4769 (1986).

19. Strehler, B. J. "Further Studies on the Thermal Induced Aging of *Drosophila melanogaster,*" *J. Gerontol.* 17:347 (1962).

20. Lindquist, S. "The Heat Shock Response," *Ann. Rev. Biochem.* 55:1151–1191 (1986).

21. Petersen, N. S. "Effects of Heat and Chemical Stress on Development," *Advances in Genetics* 28:275–296 (1990).

22. Li, G. C. "Induction of Thermotolerance and Enhanced Heat Shock Protein Synthesis in Chinese Hamster Fibroblasts, Sodium Arsenite and Ethanol," *J. Cell Physiol.* 115:116–122 (1983).

23. Samson, L., and Cairns. "A New Pathway for DNA Repair in *E. coli,*" *Nature* 267:281 (1977).

24. Ritossa, F. M. "A New Puffing Pattern Induced by Heat Shock and DNP in *Drosophila,*" *Experientia* 18:571–573 (1961).

25. Lindquist, S., and E. A. Craig. "The Heat Shock Proteins," *Ann. Rev. Genet.* 22:631–677 (1988).

26. Craig, E., P. J. Kang, and W. A. Boorstein. "A Review of the Role of 70 kDa

Heat Shock Proteins in Protein Translocation across Membranes," *Antonie van Leeuwenhoek* 58:137–146 (1990).

27. Riabowol, K. T., L. A. Mizzen, and W. J. Welch. "Heat Shock Is Lethal to Fibroblasts Microinjected with Antibodies against HSP 70," *Science* 243:433–436 (1988).

28. Hallberg, R. I., K. W. Kraus, and E. M. Hallberg. "Induction of Acquired Thermotolerance in *Tetrahymena thermophila* Can Be Achieved without the Prior Synthesis of Heat Shock Proteins," *Mol. Cell Biol.* 5:2061–2070.

29. Hendrey, J., and I. Kola. "Thermolability of Mouse Oocytes Is Due to the Lack of Expression and/or Inducibility of Hsp 70," *Mol. Reproduc. Devel.* 28:1–8 (1991).

30. Tuite, M. F., N. J. Bentley, and Bossier. "The Structure and Function of Small Heat Shock Proteins: Analysis of the *Saccharomyces cerevisiae* Hsp 26 Protein," *Antonie van Leeuwenhoek* 58:147–154 (1990).

31. Jentch, S., W. Seufert, and T. Sommer. "Ubiquitin-Conjugating Enzymes: Novel Regulators of Eukaryotic Cells," *TIBS* 15:195–198 (1990).

32. Dice, J. F., and S. A. Goff. "Error Catastrophe and Aging: Future Directions of Research," in *Modern Biological Theories of Aging*, H. R. Warner, R. N. Butler, R. L. Sprott, and E. L. Schneider, Eds. (New York: Raven Press, 1987), pp. 155–168.

33. Finley, D., B. Bartel, and A. Varshavsky. "The Tails of Ubiquitin Precursors Are Ribosomal Proteins Whose Fusion to Ubiquitin Facilitates Ribosome Biogenesis," *Nature* 338:394–401 (1989).

34. Picologou, S., N. Brown, and S. W. Liebman. "Mutations in RAD6, a Yeast Gene Encoding a Ubiquitin-Conjugating Enzyme, Stimulate Retrotransposition," *Mol. Cell Biol.* 10(3):1017–1022 (1990).

35. Pratt, G., Q. Deveraux, and M. Rechsteiner. "Ubiquitin Metabolism in Stressed Cells," in *Stress-Induced Proteins*, M. L. Pardue, J. R. Feramisco, S. Lindquist, Eds., UCLA Symposium on Molecular and Cellular Biology 96:149–162 (1989).

36. Bonner, W. M. "Metabolism of Ubiquitinated H2A," in *The Ubiquitin System*, M. Schlesinger and A. Hershko, Eds. (Cold Spring Harbor, NY: Cold Spring Harbor Laboratory, 1988), pp. 155–158.

37. Davie, J. R., S. E. Nickel, and J. A. Ridsdale. "Ubiquitinated Histone H2B Is Preferentially Located in Transcriptionally Active Chromatin," in *The Ubiquitin System*, M. Schlesinger and A. Hershko, Eds. (Cold Spring Harbor, NY: Cold Spring Harbor Laboratory, 1988), pp. 159–163.

38. Desrosiers, R., and R. M. Tanguay. "Methylation Histones at Proline, Lysine, and Arginine Residues During Heat Shock," *J. Biol. Chem.* 4686–4692.

39. Higgins, C. F. "DNA Supercoiling, Chromatin Structure and the Regulation of Gene Expression," *Antonie van Leeuwenhoek* 58:51–55 (1990).

40. Carper, S. W., P. M. Harari, and D. J. M. Fuller. "Biochemical and Cellular Response to Hyperthermia in Cancer Therapy," in *Stress-Induced Proteins*, M. L. Pardue, J. R. Feramisco, S. Lindquist, Eds., UCLA Symposium on Molecular and Cellular Biology 96:247–256 (1989).

41. Hahn, G. M., M. K. I. Adwankar, and V. S. Basrur. "Survival of Cells Exposed to Anticancer Drugs after Stress," in *Stress-Induced Proteins*, M. L. Pardue, J. R. Feramisco, S. Lindquist, Eds., UCLA Symposium on Molecular and Cellular Biology 96:223–234 (1989).

42. Lepock, J. R., K. H. Cheng, and J. Kruuv. "Thermotropic Lipid and Protein Transitions in Chinese Hamster Lung Cell Membranes: Relationship to Hyperthermic Cell Killing," *Can. J. Biochem. Cell Biol.* 61:421–427 (1983).

43. Yatvin, M. B. "The Influence of Membrane Lipid Composition and Procaine on Hyperthermic Death of Cells," *Int. J. Radiat. Biol.* 32:513–521 (1977).

44. Witkin, E. M. "Ultraviolet Mutagenesis and Inducible DNA Repair in *Escherichia coli*," *Bacteriol. Rev.* 40:869–907.

45. Walker, G. C. "Mutagenesis and Inducible Responses to Deoxyribonucleic Acid Damage in *Escherichia coli*," 48:60–93 (1984).

46. Storz, G., L. A. Tartaglia, and B. N. Ames. "The OxyR Regulon," *Antonie van Leeuwenhoek* 58:157–161 (1990).

47. Bhagwat, A. A., and S. K. Apte. "Comparative Analysis of Proteins Induced by Heat Shock, Salinity and Osmotic Stress in the Nitrogen-Fixing Cyanobacterium *Anabaena* sp. Strain L31," *J. Bacteriology* 171:5187–5189 (1989).

48. Elespuru, R. K. "Inducible Responses to DNA Damage in Bacteria and Mammalian Cells," *Environ. Mol. Mutagen.* 10:97–116 (1987).

49. Miskin, R., and E. Reich. "Plasminogen Activator: Induction of Synthesis by DNA Damage," *Cell* 19:217–224 (1980).

50. Ueda, K., and O. Hayaishi. "ADP-Ribosylation," *Ann. Rev. Biochem.* 54:73–100 (1985).

51. Mezzina, M., and S. Nocentini. "DNA Ligase Activity in UV-Irradiated Monkey Kidney Cells," *Nucl. Acids Res.* 5:4317–4334 (1978).

52. Sarasin, A. "SOS Response in Mammalian Cells," *Cancer Invest.* 3(2):163–174 (1985).

53. Herrlich, P., P. Angel, and H. J. Rahmsdorf. "The Mammalian Genetic Stress Response," *Adv. Enzyme Regul.* 25:485–504 (1986).

54. Herrlich, P., U. Mallick, and H. Ponta. "Genetic Changes in Mammalian Cells Reminiscent of an SOS Response," *Hum. Genetics* 67:360–368 (1984).

55. Lieberman, M. W., L. R. Beach, and R. D. Palmiter. "Ultraviolet Radiation Induced Metallothionein-I Gene Activation Is Associated with Extensive DNA Demethylation," *Cell* 35:207–214 (1983).

56. Rahmsdorf, H. J., N. Koch, and U. Mallick. "Regulation of Class I Invariant Chain Expression: Induction of Synthesis in Human and Murine Psmocytoma Cells by Arresting Replication," *EMBO J.* 2:811–816 (1983).

57. Schorpp, M., U. Mallick, and H. J. Rahmsdorf. "UV-Induced Extracellular Factor from Human Fibroblasts Communicates the UV Response to Nonirradiated Cells," *Cell* 37:861–868 (1984).

58. Emerit, I., and P. A. Cerruti. "Tumor Promoter Phorbol 12-Myristate 13-Acetate Inducts Clastogenic Factor in Human Lymphocytes," *Proc. Nat. Acad. Sci.* 79:7509–7513 (1982).

59. Ronai, Z. A., E. Okin, and I. B. Weistein. "Ultraviolet Light Induces the Expression of Oncogenes in Rat Fibroblast and Human Keratinocyte Cells," *Oncogene* 2:201–204 (1988).

60. Maltzman, W., and L. Czyzyk. "UV Irradiation Stimulates Levels of p53 Cellular Tumor Antigen in Nontransformed Mouse Cells," *Mol. Cell. Biol.* 4:1689–1694 (1984).

61. Williams, T. J. "Determination of Clonal Life Span in *Paramecium*," PhD Thesis, University of Wyoming, Laramie, WY (1980).

62. Keiding, J., and O. Westergaard. "Induction of DNA Polymerase Activity in Irradiated *Tetrahymena* Cells," *Exp. Cell Res.* 64:317–322 (1971).
63. Wilson, V. L., and P. A. Jones. "Inhibition of DNA Methylation by Chemical Carcinogens in Vitro," *Cell* 32:239–246 (1983).
64. Rothman, J. E. "Polypeptide Chain Binding Proteins: Catalysts of Protein Folding and Related Processes in Cells," *Cell* 59:591–601 (1989).
65. Dice, J. F. "Lysosomal Pathways of Protein Degradation," in *The Ubiquitin System*, M. Schlesinger and A. Hershko, Eds. (Cold Spring Harbor: Cold Spring Harbor Laboratory, 1988), pp. 141–146.
66. Yost, H. J., and S. Linquist. "RNA Splicing Is Interrupted by Heat Shock and Is Rescued by Heat Shock Protein Synthesis," *Cell* 45:185–193.
67. Yost, H. J., and S. Lindquist. "Translation of Unspliced Transcripts after Heat Shock," *Science* 242:1544–1548.
68. Buzin and N. Bournias-Vardiabasis. "Teratogens Produce a Subset of Small Heat Shock Proteins in *Drosophila* Primary Embryonic Cell Cultures," *Proc. Nat. Acad. Sci.* 81:4075–4079 (1984).
69. Friedberg, E. C. "Deoxyribonucleic Acid Repair in the Yeast *Saccharomyces cerevisiae*," *Microbiol. Rev.* 52:70–102 (1988).
70. Bournias-Vardiabasis, N., R. L. Teplitz, G. F. Chernoff, and R. L. Seecof. "Detection of Teratogens in *Drosophila* Embryonic Cell Culture Test Assay of 100 Chemicals," *Teratology* 28:109–122 (1983).
71. Tregar, J. M., K. A. Heichman, and K. McEntee. "Expression of the Yeast UB14 Gene Increases in Response to DNA-Damaging Agents and in Meiosis," *Mol. Cell. Biol.* 8:1132–1136 (1988).
72. McClanahan, T., and K. McEntree. "DNA Damage and Heat Shock Dually Regulate Genes in *Saccharomyces cerevisiae*," *Mol. Cell. Biol.* 6:90–96 (1986).
73. Higo, H., K. Higo, and J. Y. Lee. "Effects of Exposing Rat Embryos in Utero to Physical or Chemical Teratogens Are Expressed Later as Enhanced Induction of Heat Shock Proteins When Embryonic Hearts Are Cultivated In Vitro," *Teratogenesis, Carcinogenesis, and Mutagenesis* 8:315–328 (1988).
74. Kingston, R. E., A. S. Baldwin, Jr., and P. A. Sharp. "Regulation of Heat Shock Protein 70 Gene Expression by c-myc," *Nature* 12:280–282 (1984).
75. Fornace, A., I. Alamo, and C. M. Hollander. "Induction of Heat Shock Protein Transcripts and B2 Transcripts by Various Stresses in Chinese Hamster Cells," *Exp. Cell Res.* 182:61–74 (1989).
76. Fornace, A., Jr., A. Isaac, Jr., and C. Hollander. "Ubiquitin mRNA is a Major Stress-Induced Transcript in Mammalian Cells," *Nucl. Acid Res.* 17:1215–1229 (1989).
77. Schardein, J. L. *Chemically Induced Birth Defects* (New York: Dekker, 1985).
78. Petersen, N. S., and H. K. Mitchell. "Effects of Heat Shock on Gene Expression During Development: Induction and Prevention of the Multihair Phenocopy in *Drosophila*," in *Heat Shock from Bacteria to Man*, J. Schlesinger, M. Ashburner, and A. Tissieres, Eds. (Cold Spring Harbor, NY: Cold Spring Harbor Laboratory, 1982), pp. 345–352.
79. Mitchell, H. K., J. Roach, and N. Petersen. "Morphogenesis of Cell Hairs in *Drosophila*," *Devel. Biol.* 95:387–398 (1983).

CHAPTER 3

DNA Repair: As Influenced by Age, Nutrition, and Exposure to Toxic Substances

Ronald Hart, Ming Chou, Ritchie Feuers, Julian Leakey, Peter Duffy,
Beverly Lyn-Cook, Jack Lipman, Kenji Nakamura, Angelo Turturro,
and William Allaben, National Center for Toxicological Research,
Jefferson, Arkansas

INTRODUCTION

A critical component in assessing the effect of low-level exposure to toxicants is the repair of a genome that may be damaged by the agent. Because the adverse effect of many low-level toxicants is below practical observable thresholds, models are necessary to extrapolate the observable effects to realistic levels.[1] In these models, those agents that work through damaging the genome are usually considered to have irreversible effects — often a consequence of the procedure, which extrapolates from doses of agent that can often overwhelm cellular defenses and cause some toxic damage.[2] At low levels of agent, the normal processes of cellular defense and protection of genomic integrity are likely to be intact, influencing the response to a toxin. One of the most important factors in the protection of genomic integrity is DNA repair. The term actually covers a multitude of activities. We will concentrate on excision repair, which directly reverses the effects of a toxicant binding to the DNA.

Just as DNA repair modulates the response of a genome to an insult, DNA repair is itself modulated by a number of environmental factors. Among the important modulators of DNA repair are age, nutrition, and toxicants themselves.

AGE

Although there have been many attempts to address the effect of aging on different aspects of DNA repair, many studies are difficult to interpret. Some use cells kept in vitro (with questionable relevance to in vivo aging);

Table 3.1. Age- and UV-Induced Repair in Rats

Age (Months)	Hepatocytes		Kidney Cells	
	Ad Lib	Restricted	Ad Lib	Restricted
5	5.7	a	a	a
13	4.9	5.7	a	a
22	3.1	4.0	2.0	2.4
28	2.7	3.3	1.3	1.9
34	a	3.0	a	1.3

Source: Data from Weraarchakul et al.[14]

Note: F-344 male rats; restricted are fed 60% of ad lib diet. All values are ratios of dpm/μg DNA or radiated/irradiated cells after 1 hr. All ratios are significantly different (p < 0.01). Hepatocyes irradiated with 877 J/m^2, kidney cells with 100 J/m^2.
[a]Not done.

others compare cells from newborns to adults (confounding aging with development); and many studies fail to adequately control differences in cell replication,[3] although it was shown clearly over 15 years ago that this was an critical variable.[4] However, some studies are useful to discuss.

Rat retinal ganglion cells treated in culture, from different aged animals, demonstrated little change with age in the capacity to repair damage.[5] Lens epithelial cells also showed little change with age.[6] Various older studies using lymphocytes have given contradictory results; however, wide variability is one of the most salient characteristics.[7-10] Recent studies with better controls have been more definitive. Human lymphocytes do not seem to exhibit an overall decrease in repair capability after maturity.[11] However, a more detailed study has shown that a small subset of aging cells, in fact, lose their capability for repair of X-ray damage.[12] Skin fibroblasts do not seem to have major age-related changes.[13] The situation is quite different in hepatocytes and kidney cells.[14] In those cells, there appears to be a definite age-related decline in UV repair with age (Table 3.1).

Another approach to the relationship of aging and DNA repair compares repair across species with differing life spans. Starting with Hart and Setlow,[15] a number of other studies have shown a good correlation between life span and UV repair capacity in skin fibroblasts.[3]

One complicating factor in these studies, which has not been addressed at all, is the circadian rhythm of DNA repair. This is illustrated in Table 3.2, in which significantly different levels of DNA repair (O^6-methylguanine

Table 3.2. Circadian Variation in O^6-Methylguanine Repair

Time	Activity
1	0.19
12	0.34
20	0.32

Source: Lipman et al.[19]

Note: Skin cells from 30-month-old B6C3F1 mice. Activity is O^6-methylguanine acceptor protein activity in pmols/mg DNA. Time is hours after lights on.

Table 3.3. DNA Repair and Caloric Restriction

Damaging Agent	Brown-Norway		BNXF-344		B6C3F1	
	Ad Lib	Restricted	Ad Lib	Restricted	Ad Lib	Restricted
MMS	1.16	1.18	1.40	1.61	a	a
UV	1.38	1.41	2.07	2.75	a	a
(MGAP)	0.38	0.65	a	a	0.34	0.46

Source: Lipman et al.[19]

Note: MMS is at 0.5 mmM, UV is 20 J/m². Skin cells from rat (Brown-Norway and cross with F-344), and mouse (B6C3F1; 12 hours after lights on); restricted are fed 60% of ad lib. Values are ratios of stimulated/unstimulated cells for damaging agents. MGAP is activity of O^6-methylguanine acceptor protein, an index of capacity for repair, in pmols/mg DNA.

[a]Not done.

repair) are seen at different times of day. Different physiological states, such as estrous, can also significantly alter DNA repair.[16,17] This isozyme is male specific, and is responsible for the metabolism of a number of agents, such as doxylamine and aflatoxin B1 (AFB1).

It is instructive to look at AFB1 in a little more detail. As a result of changes in agent metabolism, the urinary excretion of AFB1 increases 40%, its plasma clearance increases by 70%, and its hydrophilic metabolites more than double.[18] This increased hydrophilicity results in increased excretion. The net effect is to reduce the binding of an equivalent dose of AFB1 to the liver genome to more than half (from 39.5 to 15.2 pmol AFB1/mg DNA, a 66% decrease), with a 40% CR.

Directly measuring DNA repair in cells isolated from CR animals shows that DNA repair increases in the rat and mouse with CR (Table 3.3). This is true for both forms of excision repair, as illustrated by the effects on damage induction by methylmethanesulfonate (MMS) and ultraviolet light. In addition, O^6-methylguanine acceptor protein, which is a repair system for a particular damage, is also elevated.[19]

Finally, CR can protect the genome in a number of different ways. Inappropriate expression of oncogenes has been shown to be associated with cancer.[20] Lyn-Cook and Hass have shown that CR results in the hypermethylation of an *HA-ras* oncogene.[21] Hypermethylation is usually associated with turning off of gene function,[22] suggesting that potentially damaging genes can be turned off by CR.

By approaching DNA repair as part of a larger process of protecting genomic integrity, it can be seen that nutrition can affect almost every aspect associated with genomic integrity, including DNA repair itself.

TOXICANTS

The best characterized system used to understand the effects of toxicants in modulating DNA repair is the adaptive response, seen in *E. coli* and mammalian cells.[23] Recent advances in the area (the development of cDNA for the protein) are likely to bring rapid further understanding in this area.[24] It is clear that there is an alkyltransferase activity in rat liver, which is inducible by alkylating agents.[25] This activity may play an important part in the removal of O^6-methylguanine after chronic exposure.[26] The effect may be part of a nonspecific tissue response. The only liver cells that demonstrate the response are parenchymal cells,[27] and the level of response, which is less than 10-fold in rat (compared to over 100-fold in bacteria). There is not a similar induction in any other system, although ionizing radiation induces the alkyltransferase in different systems.[27] The situation in humans is unclear.[27]

It is clear that toxicants can stimulate repair. This is important in understanding how toxicants will interact in inducing damage; it is especially germane to the problems of assessing human risk to low levels of agents because agents are almost never presented alone, but in mixtures with other toxicants. At the level of genomic damage, the focus has been on whether DNA-damaging agents are additive or synergistic. But evidence is building that antagonism must also be considered, as it is for metabolism and other factors in human risk.[1] The relevance of this consideration to hormesis is especially interesting.[28] In certain situations, exposure to an agent may predispose cells to be less sensitive to other agents — perhaps especially true for low levels of radiation.

CONCLUSION

In evaluating the risk associated with low levels of exposure to toxicants, it is clear that DNA repair, one of the main defenses against agent damage, is not a constant. It can be modified by age, time of day, and physiological state. Nutrition, especially CR, can modify almost every step in the process of protecting genomic integrity. And history of exposure can modify DNA repair. Thus, the conditions of exposure are almost as important to toxicity as the exposure level itself, even at the level of DNA repair. Extrapolation from high to low dose, to be consistent with what is known, should be less a mathematical exercise (as it is at present) than a exercise in toxicological judgment, which puts the exposure in proper perspective. This appears to be true at almost every level in the process inducing a response with atoxic stimulus, even those often thought to be very basic, such as DNA repair.

REFERENCES

1. Interagency Staff Group. "Chemical Carcinogens: A Review of the Science and Associated Principles," *Environ. Health Perspect.* 67:201–282 (1986).
2. Ames, B., and L. Gold. "Too Many Rodent Carcinogens: Mitogenesis Increases Mutagenesis," *Science* 249:970–971 (1990).
3. Turturro, A., and R. Hart. "DNA Repair Mechanisms in Aging," in *Compara-*

tive Pathobiology of Major Age-Related Diseases, D. Scarpelli and G. Migaki, Eds. (New York: A. R. Liss, 1984), pp. 19–46.

4. Hart, R., and R. Setlow. "DNA-Repair in Late-Passage Human Cells," *Mech. Age Develop.* 5:67–75 (1976).

5. Ishikawa, T., S. Takayama, and T. Kitagawa. "DNA Repair Synthesis in Rat Retinal Ganglion Cells Treated with Chemical Carcinogens or Ultraviolet Light *in vitro,* with Special Reference to Aging and Repair Level," *J. Nat. Cancer Inst.* 61:1101–1105 (1978).

6. Treton, J., and Y. Courtois. "Correlation between DNA Excision Repair and Mammalian Lifespan," *Cell Biol. Int. Rep.* 6:253–260 (1982).

7. Pero, R., C. Bryngelsson, F. Mitelman, R. Kornfalt, J. Thulin, and A. Norden. "Interindividual Variation in the Responses of Cultured Human Lymphocytes to Exposure from DNA Damaging Chemical Agents," *Mutat. Res.* 53:327–341 (1978).

8. Lambert, B., U. Ringborg, and L. Skoog. "Age-Related Decrease of Ultraviolet-Light Induced DNA Repair Synthesis in Human Peripheral Leukocytes," *Cancer Res.* 39:2792–2795 (1979).

9. Lezhava, T., V. Prokofjeva, and V. Mikhelson. "Reduction in UV-Induced Unscheduled DNA Synthesis in Human Lymphocytes at an Extreme Age," *Tsitolozica* 11:1360–1363 (1979).

10. Pero, R., and C. Ostlund. "Direct Comparison in Human Resting Lymphocytes of the Inter-Individual Variations in Unscheduled DNA Synthesis Induced by N-Acetoxy-2-acetylaminofluorene and Ultraviolet Radiation," *Mutat. Res.* 73:349–361.

11. Roth, M., L. Emmons, M. Haner, H. Muller, and J. Boyle. "Age-Related Decrease in an Early Step of DNA-Repair of Normal Human Lymphocytes Exposed to Ultraviolet-Irradiation," *Exp. Cell Res.* 180:171–177 (1989).

12. Singh, N. P., D. Danner, R. Tice, L. Brant, and E. Schneider. "DNA Damage and Repair with Age in Individual Human Lymphocytes," *Mutat. Res.* 237:123–130 (1990).

13. Tice, R., and R. Setlow. "DNA Repair and Replication in Aging Organisms and Cells," in *Handbook of the Biology of Aging,* C. Finch and E. Schneider, Eds. (New York: Van Nostrand Reinhold, 1985), pp. 173–224.

14. Weraarchakul, N., R. Strong, W. Wood, and A. Richardson. "The Effect of Aging and Dietary Restriction on DNA Repair," *Exp. Cell Res.* 181:197–204 (1989).

15. Hart, R., and R. Setlow. "Correlation between Deoxyribonucleic Acid Excision-Repair and Life-Span in a Number of Mammalian Species," *Proc. Nat. Acad. Sci.* 71:2169–2173.

16. Braun, R., T. Ratko, J. Pezzuto, and C. Beattie. "Estrous Cycle Modification of Rat Uterine DNA Alkylation by N-Methylnitrosourea," *Cancer Lett.* 37:345–352.

17. Ratko, T., R. Braun, J. Pezzuto, and C. Beattie. "Estrous Cycle Modification of Rat Mammary Gland DNA Alkylation by N-Methyl-N-nitrosourea" *Cancer Res.* 48:3090–3093.

18. Pegram, R., W. Allaben, and M. Chou. "Effect of Caloric Restriction on Aflatoxin B_1-DNA Adduct Formation and Associated Factors in Fischer 344 Rats: Preliminary Findings," *Mech. Age Develop.* 48:167–177 (1989).

19. Lipman, J., A. Turturro, and R. Hart. "The Influence of Dietary Restriction on

DNA Repair in Rodents: A Preliminary Study," *Mech. Age Develop.* 48:135–143 (1989).

20. Travali, S., J. Koniecki, S. Petralia, and R. Baserga. "Oncogenes in Growth and Development," *FASEB J.* 4:3209–3214 (1990).

21. Hass, B., B. Lyn-Cook, L. Poirier, and R. Hart. "Differences in Growth Rate and DNA Methylation and in *c-Ha-ras* in Pancreatic Acinar Cells of Brown-Norway Rats Fed *ad libitum* and Caloric Restricted Diets," *Proc. Amer. Assoc. Cancer Res.* 32:862 (1991).

22. Poirier, L., W. Zapisek, and B. Lyn-Cook. "Physiological Methylation in Carcinogenesis," in *Mutation and Environment. Part D* (New York: Wiley-Liss, 1991), pp. 97–112.

23. Lindahl, T., B. Sedgwick, M. Sekiguchi, and U. Nakabeppu. "Regulation and Expression of the Adaptive Response to Alkylating Agents," *Ann. Rev. Biochem.* 57:133–157.

24. Tano, K., S. Shiota, J. Collier, R. Foote, and S. Mitra. "Isolation and Structural Characterization of a cDNA Clone Encoding the Human DNA Repair Protein for O^6-Alkylguanine," *Proc. Nat. Acad. Sci.* 87:686–690 (1990).

25. Saffhill, R., G. Margison, and P. O'Conner. "Mechanisms of Carcinogenesis Induced by Alkylating Agents," *Biophys. Acta* 823:111–145 (1985).

26. Swenburg, J., M. Dyroff, J. Bedell, J. Popp, N. Huh, U. Kirstein, and M. Rajewsky. "O^4-Ethyldeoxythymidine, But Not O^6-Methyldeoxyguanosine, Accumulates in Hepatocyte DNA of Rats Exposed Continually to Diethylnitrosamine," *Proc. Nat. Acad. Sci.* 81:1692–1695 (1984).

27. Pegg, A. "Properties of Mammalian O^6-Alkylguanine-DNA Transferases," *Mutat. Res.* 233:165–175 (1990).

28. Boxenbaum, H. "Assessing the Nature and Quality of the Hormesis Database," this volume.

CHAPTER 4

Biochemical Mechanisms of Biphasic Dose-Response Relationships: Role of Hormesis

Harihara M. Mehendale, Department of Pharmacology and Toxicology, University of Mississippi Medical Center, Jackson, Mississippi

INTRODUCTION

A fundamental principle underlying the toxic actions of chemical and physical agents is that greater toxicities are manifested with increasing doses.[1] With the exception of cancer and certain immune reactions, it is generally recognized that there is a threshold dose before which toxic manifestations do not begin to appear. Living organisms have a remarkable capability to overcome injury from physical or chemical agents encountered in their environment. In an attempt to enhance survival from many kinds of noxious injuries, organisms have developed several kinds of cellular and tissue defense mechanisms. While unicellular organisms may rely on cellular defense mechanisms for survival, multicellular organisms have developed additional integrated defense mechanisms through the evolved sophistication of tissues and organs, with matching structural and functional complexity. Generally speaking, such mechanisms might be classified into two major categories, and each category might be viewed as a component of a two-tier defense system. One category is represented by biochemical mechanisms that enable the organism to prevent infliction of injury after a noxious insult. The second is a biological response intended to overcome injury, by promoting tissue healing after the fact. Although much attention has been focused on the endogenous biochemical defense mechanisms that participate in preventing the infliction of cellular and tissue injury by directly or indirectly interfering with the inflictive mechanisms, the biology of endogenous mechanisms that might be recruited to overcome tissue injury after it occurs has received little attention.

The threshold for toxic actions of a toxic chemical at the first tier might be defined by the extent to which the cellular or tissue defense mechanisms might be able to prevent infliction of injury. An example of this might be illustrated by the depletion of intracellular glutathione by the acetaminophen metabolite, acetamido iminoquinone. Since actual infliction of cellular injury has been attributed to the formation of this active metabolite in the liver, as long as cellular glutathione levels are not decreased below a critical (threshold) level, significant injury does not occur. Once the cellular glutathione level dips to the critical low level (subthreshold), continued formation of the iminoquinone metabolite would result in covalent binding of this metabolite to cellular macromolecules, resulting in greater infliction of injury and cell death. However, even at this stage, all is not lost. After the injury has occurred, the tissue is able to recruit a second tier of tissue defense mechanisms.[2-4] The dead or dying hepatocytes might be replaced by new cells via stimulated hepatocellular regeneration and tissue repair.[2] As long as this second tier mechanism itself is not interfered with, as in the case of limited tissue injury after a low but toxic dose of acetaminophen,[3] complete recovery occurs with no further adverse consequences.

These endogenous cytoprotective mechanisms, as well as the mechanisms dependent on the biology of cellular proliferation and tissue repair, are referred to as hormetic mechanisms. Higher doses of toxic chemicals may interfere with this tissue defense mechanism, resulting in greater and more accelerated permissive progression of tissue injury until the tissue is completely destroyed. Therefore, at subthreshold levels of toxic chemicals, the hormetic mechanisms are not subdued, suppressed, or interfered with in any other fashion. At doses exceeding the threshold levels, hormesis is compromised, causing the cellular and tissue injury to progress in an uncontrolled fashion, resulting in accelerated injury.

Evolution of hormesis as an adaptation mechanism in our struggle to survive adverse or noxious insult is apparent in several experimental as well as ambient settings. Resistance to the cytotoxic actions of low-level radiation after repeated exposure to even lower levels of radiation has a mechanistic basis in hormesis.[5,7] Recent evidence also points to activation of hormetic mechanisms against cytotoxic injury from other free-radical generating mechanisms.[8-10] It was anticipated from studies in which microsomal incubations of menadione, a redox cycling quinone, were employed that exposure to repeated phenobarbital administration would result in greater cytotoxicity to isolated hepatocytes.[11] Actual incubations of hepatocytes isolated from rats preexposed to phenobarbital with menadione revealed no such increase in toxicity.[8] Additional experimental inquiry revealed that several cytoprotective mechanisms were also induced simultaneously as the free-radical generating potential was also increased.[8-10]

The objective of this chapter is to discuss the role of hormetic mechanisms in determining the final outcome of toxicity at low versus high doses of chemicals. The cytoprotective defense mechanisms, along with inducible hormetic mechanisms, appear to work in tandem to overcome toxicities associated with low-level exposures to toxic chemicals. Interference with these mechanisms may result in remarkable progression of tissue injury in an unabated fashion. A greater understanding of the biochemical mechanisms underlying hormesis would not only lead to a better understanding of the mechanistic basis for the "threshold" concept, but it is also likely to provide us with a rational basis for the assessment of risk at low-level exposures to chemical and physical agents.

This chapter focuses on the hepatotoxic effects of low-level exposure to halomethane solvents. The discovery of tissue-healing mechanisms stimulated as a hormetic response to limited liver injury was made only through work with exposures of experimental animals to low levels of halomethanes. The most compelling reason for investigating the toxicology of low levels of halomethanes was provided by the observation that at individually nontoxic doses, the combination of exposure to a chlorinated pesticide chlordecone (Kepone) and a halomethane such as CCl_4 causes an unprecedented level of toxic injury.[2] In order that the experimental evidence for the existence of inducible hormetic mechanisms in the form of tissue repair directed to overcoming tissue injury can be discussed in the context of its discovery, the toxicology of halomethanes, and the interactive toxicity of halomethanes in combination with exposure to other chemicals, will be reviewed.

MECHANISM OF HALOMETHANE HEPATOTOXICITY

Carbon Tetrachloride

The mechanism of CCl_4-induced hepatotoxicity has been extensively studied.[12-23] Since the mechanism underlying the toxicology of CCl_4 is central to the consideration of how its toxicity might affect the liver tissue and how this might be modified by other chemicals, it is worthwhile to outline the prevailing concepts concerning the hepatotoxicity of CCl_4.

Several reviews have appeared on this topic.[12-16] The leading theory for the mechanism of cellular damage caused by CCl_4 is that the compound is bioactivated by cytochrome P-450 mediated reactions to $\cdot CCl_3$ free-radical,[12-15,17-23] which is further converted to a peroxy radical, $CCl_3O\cdot_2$.[15,20] There is evidence for covalent binding of CCl_4 upon bioactivation.[12-23] The $CCl_3O\cdot_2$ radical is also thought to decompose to phosgene and electrophilic Cl^-, which can react with other macromolecules.[24] The free radicals $\cdot CCl_3$ and $CCl_3O\cdot_2$ readily react with polyunsaturated fatty acids of the endoplasmic reticulum and other hepatocellular membranes to initiate the formation of organic lipid peroxides. In the presence of cellular O_2, these organic peroxy radicals in turn can react with other polyunsaturated fatty acids to perpetuate a series of self-propagating chain reactions, a process commonly referred to as "propagation of lipid peroxidation."[12] The bioactivation of CCl_4 and initiation of the self-propagating lipid peroxidation, working in tandem, destroy the cellular membranes, leading to cell death. The principal hepatic lesion is characterized by centrilobular necrosis,[25] the extent of injury depending upon the dose. Demonstration of the metabolism of CCl_4 to $CHCl_3$ and to CO_2,[26,27] and covalent binding of CCl_4 to liver protein and lipid,[19,27] lend experimental support to the bioactivation theory of CCl_4 injury.[15-28]

Bromotrichloromethane and Chloroform Toxicity

Hepatotoxicity of $BrCCl_3$, a brominated analog of CCl_4, is also due to its metabolism to the same $\cdot CCl_3$ radical formed from CCl_4.[18,29-31] Much greater toxicity of this compound[29,30] in comparison to CCl_4 has been attributed to the relative ease with which the C-Br bond can be cleaved.[18] A clear inverse relationship exists between the bond dissociation energy of this series of halomethanes ($BrCCl_3$ < CCl_4 < $FCCl_3$ < $HCCl_3$) and their potency to initiate free-radical reactions,[12-16] to produce lipid peroxidation, and to produce liver necrosis.

With regard to $CHCl_3$, several investigations suggest that phosgene, a reactive metabolite of $CHCl_3$, is responsible for its hepatotoxic,[32,33] nephrotoxic,[34] and possibly its carcinogenic[32,35] effects. Hepatotoxic effects are due to phosgene-mediated cellular glutathione (GSH) depletion in tandem with the increased covalent binding to hepatocellular macromolecules.[32,36] Although, like CCl_4, $CHCl_3$ also needs metabolic activation to exert its full necrogenic potential, unlike CCl_4, lipid peroxidation is not involved in hepatocellular necrosis. Recently, however, lipid peroxidation has been claimed in $CHCl_3$ toxicity.[37,38] A second important distinction is that, unlike CCl_4, metabolism of $CHCl_3$ to a free-radical form has not been associated with its necrogenic action.[32] Recent studies on $CHCl_3$ toxicity have involved mouse hepatocyte primary cultures[39] and Mongolian gerbils.[40-42]

Mechanism of CCl_4-Autoprotection

A small dose of CCl_4 is known to protect against a subsequently administered large dose of CCl_4.[43-46] Several lines of evidence have accumulated[43-46] to establish that the mechanism of this "autoprotection" is related to the destruction of liver microsomal cytochrome P-450 by the initial protecting dose of CCl_4. Reports[12,13,19,47] demonstrating the destruction of a specific form of cytochrome P-450 provide additional support for $\cdot CCl_3$ free-radical mediated destruction of cytochrome P-450. This presumably results in compromised bioactivation of a subsequently administered large dose of CCl_4. Since bioactivation of CCl_4 is an obligatory step for CCl_4 injury, the subsequently administered large dose of CCl_4 cannot inflict massive liver injury as would be ordinarily expected.[12,13] Recent time-course studies have revealed that actual liver injury sustained by the autoprotected animal is indistinguishable from that sustained by an animal not receiving the protective dose of CCl_4. These findings have prompted a search for an alternate mechanism for the phenomenon of autoprotection.[48] The protective dose-stimulated hepatocellular regeneration and tissue repair enable the autoprotected animal to recover from the same level of massive injury.[48] Nonetheless, the mechanism of cytochrome P-450 destruction by CCl_4 is of interest. Although a direct demonstration of $\cdot CCl_3$ and $CCl_3O\cdot_2$ radicals has been claimed using spin trapping techniques,[13,20,49,50] the precise events leading to the destruction of cytochrome P-450 by CCl_4 remain elusive. One view holds that the $\cdot CCl_3$ free radical directly interacts with the endoplasmic reticulum and destroys the cytochrome P-450;[47,49,50] another view holds that lipid peroxidation initiated by the $\cdot CCl_3$ free radical results in the destruction of cytochrome P-450.[51] There is evidence that the lipid peroxidation initiated by the $\cdot CCl_3$ or $CCl_3O\cdot_2$ radicals results in the release of 4-hydroxynonenol,[52] which has been demonstrated to inhibit cytochrome P-450 mediated mixed-function oxygenase (MFO) activity.[53] The demonstration[28,54] that phosgene is a metabolite of CCl_4 has raised the possibility that this reactive metabolite may also be involved in the destruction of cytochrome P-450.

POTENTIATION OF HALOMETHANE HEPATOTOXICITY

The enhancement of CCl_4 toxicity by a variety of chemicals has been observed: phenobarbital,[55-58] aliphatic alcohols,[59,60] ketones,[61,62] DDT,[63] polychlorinated biphenyls,[64] and other experimental manipulations.[65-68] Pyrazole,[69] vitamin A,[70] other halomethanes,[71] and complex chemical waste mixtures[72] are representative examples. 3-Methylcholanthrene protects against CCl_4 hepatotoxicity.[73,74] Treatment with cysteamine, cysteine, or SKF-525A[75,76] — and a number of other chemicals[77-80] — afford protection against CCl_4 hepatotoxicity. Studies also indicate protection by partial hepatectomy,[81-86] and by an externally supplemented source of energy.[87-90] Most, if not all, experimental conditions that potentiate the toxicity of CCl_4 correlate with increased hepatic microsomal cytochrome P-450 content and with accordingly increased bioactivation of CCl_4 in the liver. Hepatocellular injury is increased as a consequence of the enhanced production of free-radical forms of CCl_4 metabolites.

Hepatotoxicity of $BrCCl_3$ is also known to be potentiated by agents known to induce drug-metabolizing enzymes of the liver.[30,91,92] Hepatotoxicity of other halomethanes related to $CHCl_3$ is also potentiated by other chemicals. Hepato- and nephrotoxicity of $CHCl_3$ is potentiated by aliphatic alcohols,[93-95] ketones,[96,97] and phenobarbital.[32]

The widely accepted theory for the mechanism of xenobiotic-induced enhancement of liver injury caused by CCl_4 is that its bioactivation to $\cdot CCl_3$ and $CCl_3O\cdot_2$ free radical is increased.[12-27] There is evidence for increased covalent binding of CCl_4 to liver tissue upon bioactivation.[12,13,19] Increased

Table 4.1. Amplification of Lethal Effects of Several Halomethanes by Dietary Exposure of Rats to Subtoxic Contaminants

Dietary Pretreatment	Halomethane	48-hr LD_{50} (mL/kg)	Increase in Toxicity (-fold)
Female rats			
Control	CCl_4	1.25	—
Chlordecone (10 ppm)	CCl_4	0.048	26
Male rats			
Control	CCl_4	2.8	—
Chlordecone (10 ppm)	CCl_4	0.042	67
Phenobarbital (225 ppm)	CCl_4	1.7	1.6[a]
Control	$BrCCl_3$	0.119	—
Chlordecone (10 ppm)	$BrCCl_3$	0.027	4.5

Source: Adapted from Mehendale.[16]

[a]Not significant at $P \leq 0.05$.

production of $\cdot CCl_3$ and $CCl_3 O\cdot_2$ radicals leads to increased lipid peroxidation, culminating in increased liver injury.[12]

Interactive Toxicity of Chlordecone and CCl_4

From a perspective of public health, a major toxicological issue is the possibility of unusual toxicity due to interaction of two or more toxic chemicals at individually harmless levels upon environmental or occupational exposures. While some laboratory models exist for such interactions for the simplest case of only two chemicals, progress in this area has suffered for want of models where the two interactants are individually nontoxic. One such model is available, where prior exposure to nontoxic levels of the pesticide Kepone (chlordecone) results in a 67-fold amplification of CCl_4 lethality in rats (Table 4.1). The mechanism of the remarkable interactive toxicity is of interest in the assessment of risk from exposure to combinations of chemicals.

Prior exposure to nontoxic level of chlordecone (10 ppm in diet for 15 days) results in a marked amplification of CCl_4 hepatotoxicity[55,98] and lethality.[56,98,99] Neither the close structural analogs of chlordecone (mirex and photomirex) nor phenobarbital (Figure 4.1) exhibit this property.[55,56] Plaa and associates[100,101] have demonstrated the capacity of chlordecone to potentiate $CHCl_3$ hepatotoxicity in mice. These observations have been extended to demonstrate that in addition to the hepatotoxic effects, lethal effect of $CHCl_3$ is also potentiated by exposure to 10-ppm dietary chlordecone[102] (Table 4.2) and that this may be associated with suppressed repair of the liver tissue.[103] Chlordecone also potentiates the hepatotoxicity and lethality of $BrCCl_3$.[91,92] While the toxicity of these closely related halomethanes is potentiated by such low levels of chlordecone (Figure 4.2), the toxicity of structurally and mechanistically dissimilar compounds (Figure

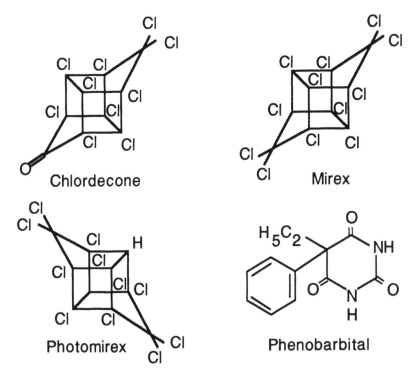

Figure 4.1. Structures of chlordecone, mirex, photomirex, and phenobarbital. Chlordecone amplifies the toxicity of several halomethanes closely related to CCl_4. Mirex and photomirex, despite being close structural analogues of chlordecone, essentially do not possess this propensity. Phenobarbital, a commonly employed drug in interaction studies at high doses, does increase liver injury of CCl_4, but this enhanced liver injury is inconsequential to animal survival and health, since the animals are able to recover from potentiated liver injury.

Table 4.2. Amplification of Lethal Effects of Halomethanes by Dietary Exposure of Mice to Subtoxic Contaminants

Dietary Pretreatment	Halomethane	48-hr LD_{50} (mL/kg)	Increase in Toxicity (-fold)
Male mice			
Control	$CHCl_3$	0.067	—
Chlordecone (10 ppm)	$CHCl_3$	0.16[a]	4.2
Mirex (10 ppm)	$CHCl_3$	0.70	no change
Phenobarbital (225 ppm)	$CHCl_3$	0.70	no change

Source: Adapted from Mehendale.[16]

[a]Significantly different at $P \leq 0.05$.

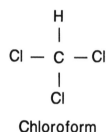

Figure 4.2. Structures of carbon tetrachloride, bromotrichloromethane, and chloroform as examples of halomethane solvents. Hepatotoxicity and lethality of these solvents is remarkably amplified by the pesticide chlordecone.

4.3, Table 4.3) is not potentiated[104] except after exposure to high levels of chlordecone.[105] This remarkable capacity to potentiate halomethane hepatotoxicity does not appear to be related to chlordecone-induced cytochrome P-450 or associated enzymes,[55,99,106] enhanced bioactivation of CCl_4,[41,42,54,86,104] increased lipid peroxidation,[55,104,105] or decreased glutathione.[107] Several candidate mechanisms were considered carefully until a novel mechanism was discovered (Table 4.4).

Mechanism of the Interactive Toxicity of Chlordecone and CCl_4

These findings led to some very basic studies concerning the progression of the hepatotoxicity during a time course following CCl_4 administration to either normal or chlordecone pretreated rats. The histochemical and histomorphometric experiments revealed that suppressed hepatocellular regeneration and tissue repair might explain the remarkable amplification of CCl_4 toxicity by prior exposure to chlordecone.[55,108,109] Similar time-course studies on Ca^{2+} levels in the liver mitochondria, microsomes, and cytosol fractions revealed a possible association of increased Ca^{2+} accumulation and suppressed hepatocellular regeneration.[110,111] Despite some reports that chlordecone interferes with Ca^{2+} uptake mechanisms in extrahepatic tissues,[112] even at toxic doses, chlordecone does not cause disruption of hepatocellular Ca^{2+},[113] while the chlordecone + CCl_4 interaction does remarkably so.[110-114]

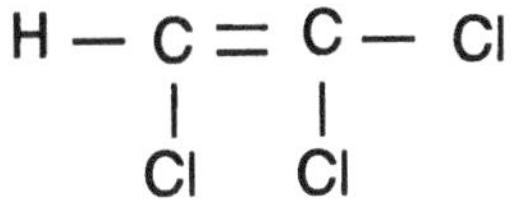

H — C = C — Cl
        | |
        Cl Cl

1,1,2-Trichloroethylene

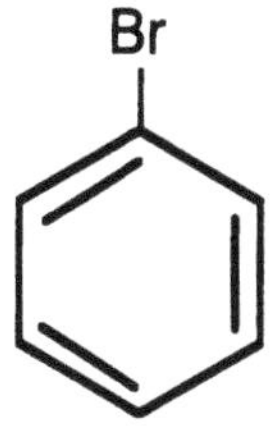

Br
⬡

Bromobenzene

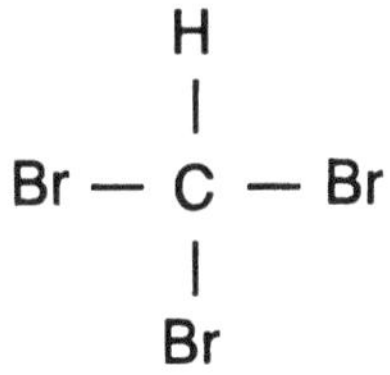

      H
      |
Br — C — Br
      |
      Br

Bromoform

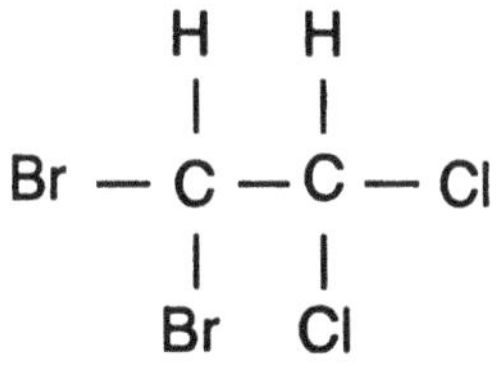

      H   H
      | |
Br — C — C — Cl
      | |
      Br Cl

Dibromodichloromethane

Figure 4.3. Structure of 1,1,2-trichloroethylene, bromobenzene, bromoform, and dibromodichloromethane. Toxicity of these chemicals is not potentiated by prior dietary exposure to 10 ppm chlordecone.

Recent studies have also shown a significant activation of phosphorylase a, a finding commensurate with the precipitous depletion of glycogen[89,90,109,115] and ATP.[89,90,115] Based on several lines of experimental evidence, a hypothesis was proposed for the mechanism of the interactive toxicity of chlordecone and CCl_4.[16]

Stimulation of Tissue Repair as a Hormetic Response to Tissue Injury

First, it became necessary to hypothesize the mechanism for why an ordinarily nontoxic dose of CCl_4 is nontoxic.[55] Figure 4.4 illustrates the mechanism of recovery from the limited liver injury observed after the

Table 4.3. Specificity of Potentiation of Halomethane Toxicity by Chlordecone

Compound	Potentiation?	References
$CHCl_3$	yes	100,102
CCl_4	yes	55,98,99
$CBrCl_3$	yes	91,92
$CHBr_3$	no	55
CBr_4	no	55
CCl_2CHCl	no	104
Bromobenzene	no	104

Table 4.4. Candidate Mechanisms of Chlordecone Amplification of Halomethane Toxicity

Mechanism	Role in Amplification
1. Enhanced bioactivation of halomethanes	Increased infliction of injury; only Stage I of toxicity is increased.
2. Increased lipid peroxidation	Not known or none
3. Estrogenic property of chlordecone	None
4. Increased Ca^{2+} accumulation; precipitous glycogenolysis, loss of ATP	Pertubed cellular biochemistry and ablation of hormetic mechanisms
5. Suppressed hepatocellular regeneration due to ablation of the early-phase hormesis	Injury progresses unabatedly. Stage II of toxicity

administration of a low dose of CCl_4 alone. Within 6 hr after the administration of a low dose of CCl_4, limited hepatocellular necrosis inflicted by the same widely accepted mechanisms of CCl_4 bioactivation followed by lipid peroxidation occurs. This limited hepatolobular injury is evident as centrilobular necrosis, ballooned cells, and steatosis. By mechanisms hitherto unexplored, simultaneously the liver tissue responds by stimulating hepatocellular regeneration.[108,109] Most interestingly, this hepatocellular division is maximal at 6 hr, even though the limited injury evident as centrilobular necrosis is only beginning to manifest at that time. Although the molecular events responsible for the stimulation of hepatocellular division have not been explored, glycogen, the principal form of hepatic energy resource, is mobilized prior to cell division.[108,109] Glycogen levels are restored after cell division has been adequately stimulated.[108] The limited hepatocellular necrosis enters the progressive phase between 6 and 12 hr,[82,85,108,109] while the hepatocellular regeneration and tissue-healing processes continue. By 24 hr, no significant liver injury is evident. These observations allow one to propose that stimulation of hepatocellular regeneration is a protective response of the liver, occurs very early after the administration of a low dose of CCl_4, and leads to replacement of dead cells, thereby restoring the hepatolobular architecture.[13,15,55]

Furthermore, this remarkable biological event results in another important protective action. It is known that newly divided liver cells are relatively resistant to toxic chemicals.[39,116-118] Therefore, in addition to the restoration of the hepatolobular architecture by cell division, due to the relatively greater resistance of the new cells, the liver tissue is able to overcome the imminence of greater injury during the progressive phase (6 to 12 hr), obtunding the spread of injury on the one hand, and speeding up the process of overall recovery through tissue healing on the other (Figure 4.4). By 6 hr, over 75% of the administered CCl_4 is eliminated in the expired air,[106] leaving less than 25% for continued injury, all of which is eliminated by 24 hr.[55] Relative resistance of the newly divided cells at this critical time

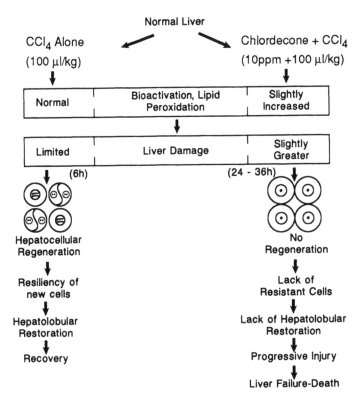

Figure 4.4. Proposed mechanism for the highly amplified interactive toxicity of chlordecone + CCl₄. The scheme depicts the concept of suppressed hepatocellular regeneration, simply permitting what is normally limited liver injury caused by a subtoxic dose of CCl_4 to progress in the absence of hepatolobular repair and healing mechanisms stimulated by the limited injury. The limited hepatotoxicity from a low dose of CCl_4 is normally controlled and held in check by the hepatocellular regeneration and hepatolobular healing. The chlordecone + CCl_4 combination treatment results in unabated progression of injury owing to a lack of tissue repair obtunded due to lack of cellular energy. These events lead to complete hepatic failure, culminating in animal death. Ongoing studies indicate that a very similar mechanism is responsible for the amplification of $CHCl_3$ and $BrCCl_3$ toxicity by chlordecone. Adapted from Mehendale[16] and Karunaratne.[25]

frame, as the animal continues to exhale the remaining CCl_4, would be an added critical defense mechanism. At later time points (12 hr and onwards), most of the CCl_4 would have been eliminated by the animal, and hence continued cellular regeneration during this time period and at later time points allows for complete restoration of the hepatolobular architecture during and after the progressive phase of injury.[82–85]

Administration of the same low dose of CCl_4 to animals maintained on food contaminated with low dose of chlordecone results in initial injury by the same mechanisms of bioactivation of CCl_4 and lipid peroxidation (Figure 4.4). The liver injury in this case is slightly greater due to the approximately doubled rate of bioactivation of CCl_4 in livers of animals preexposed to chlordecone.[15,55,106] The liver injury, thus initiated, enters the progressive phase between 6 to 12 hr, and this phase is accelerated in the absence of tissue repair mechanisms.[82-85,108,109] The highly unusual amplification of CCl_4 toxicity relates to the suppression of the initial hepatocellular regeneration, otherwise ordinarily stimulated by CCl_4 within 6 hr (Figure 4.4).

The mechanism responsible for the abrogation of this hormetic mechanism of stimulated cell division is of significant interest. At this juncture, experimental observations permit invoking a role for hepatocellular bankruptcy in cellular energy. Under conditions of increased hepatocellular injury, mobilization of hepatic glycogen is initiated in order to stimulate hepatocellular division.[109-114] Under these conditions of increased demand for cellular energy (augmented need for extrusion of extracellular Ca^{2+} from the cells, protection against free-radical mediated injury, etc.), the hepatocytes are incapacitated due to insufficient availability of cellular energy. As a result, stimulation of cell division, which normally occurs after the administration of a low dose of CCl_4, cannot occur. The failure of cell division has two important implications:

1. Hepatolobular structure cannot be restored.
2. Unavailability of newly divided, relatively resistant cells predisposes the liver to continuation of liver injury during the progressive phase (6 to 12 hr and beyond).[14-16,55,114,115]

Permissively progressive injury continues unabatedly as a consequence of the mitigated tissue repair mechanisms, leading to massive hepatic failure,[55,56,91,99] followed by animal death.[14-16]

Many studies have shown a biphasic increase in hepatocellular Ca^{2+} levels in CCl_4 toxicity.[111] The unusual aspect of excessive Ca^{2+} accumulation observed in livers treated with the chlordecone + CCl_4 combination is that it occurs at a dose of CCl_4 not ordinarily associated with the causation of increased intracellular Ca^{2+}. Furthermore, chlordecone alone, even at a 10-fold higher dose than used in the interaction studies, does not increase hepatocellular Ca^{2+}.[15,111] Although in vitro studies with cellular organelles have been employed to speculate that the failure of organelle Ca^{2+} pumps leads to increased cytosolic Ca^{2+} levels, our studies indicate that at no time did these organelles contain decreased Ca^{2+}.[15,111] Indeed, the only significant change observed with regard to organelle Ca^{2+} is increased Ca^{2+} in the organelles in association with increased liver injury.[14,111] Therefore, there is no in vivo evidence for decreased Ca^{2+} content in the organelles, which is in contradiction to the predictions from the in vitro findings.[114,115]

The primary mechanism leading to a highly amplified toxicity is a failure on the part of the biological events leading to hepatocellular division. Increased accumulation of extracellular Ca^{2+} during the progressive phase of liver injury[111] would be consistent with the significant loss of biochemical homeostasis in hepatocytes (Figure 4.4). Earlier histomorphometric[109] as well as biochemical[14,15,89,90] studies have shown that glycogen levels drop very rapidly after CCl_4 administration to chlordecone-treated animals. Increased cytosolic Ca^{2+} would be expected to result in activation of phosphorylase b to phosphorylase a, the enzyme responsible for glycogenolysis.[115] Phosphorylase a activity[114,115] and precipitous glycogenolysis[108,109,111,115] are experimental observations consistent with the rapid depletion of cellular energy on the one hand,[115] and irreversible increase in cytosolic Ca^{2+} on the other.[114]

An intriguing aspect of the experimental framework leading to the proposed mechanism is the observation that phenobarbital, even at significantly higher doses (225 ppm in the diet for 15 days), does not potentiate the lethal effect of CCl_4. Although histopathological parameters of liver injury, such as hepatocellular necrosis and ballooned cell response, are indicative of significantly enhanced hepatotoxicity by phenobarbital, if the animals are left alone, this injury does not progress to significantly increased lethality. Hepatic microsomal cytochrome P-450 is approximately doubled by prior dietary exposure to 225 ppm phenobarbital and the bioactivation of CCl_4 is tripled,[55,106] and these parameters are consistent with the enhanced initiation of liver injury measured by histopathology, elevation of serum transaminases, or by hepatic function. Nevertheless, the liver injury neither progresses in an accelerated fashion nor is irreversible, as indicated by the reversal of liver injury accompanied by animal survival.[55,56,85]

Figure 4.5 illustrates the proposed mechanism for phenobarbital-enhanced liver injury of CCl_4, which is associated with a lack of enhanced lethality. Induction of hepatomicrosomal cytochrome P-450 by phenobarbital results in approximate tripling of CCl_4 bioactivation and increased lipid peroxidation.[55,106] Enhanced liver injury is consistent with these observations (Figure 4.5). It should be recalled that the liver is normally able to respond by stimulation of hepatocellular regeneration after a low dose of CCl_4 within 6 hr (Figure 4.5). While phenobarbital exposure results in greater injury, the liver's ability to respond by stimulated cell division is not completely compromised, as evidenced by the stimulation of hepatocellular regeneration starting at 24 to 36 hr and continuing through 72 hr. Therefore, hepatocellular regeneration is stimulated, thereby counteracting the enhanced liver injury, which leads to recovery from increase in liver injury. In view of the enhanced liver injury, restoration of normal hepatolobular architecture takes longer than the approximate 24 hr required upon administration of a low dose of CCl_4 alone. Although the hepatocellular regeneration is delayed from 6 to 24 hr, when it does occur it is enhanced substantially, apparently tempered by the demand for more extensive restoration of hepatolobular architecture as a consequence of greater injury.[85,102,103] Hence, the overall effect of phenobarbital-induced potentiation of CCl_4 injury is merely to delay the stepped up hepatocellular regeneration, tissue repair,

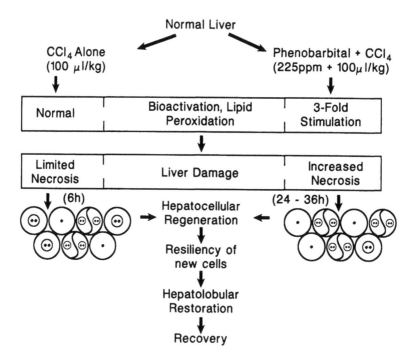

Figure 4.5. Proposed mechanism for phenobarbital-induced potentiation of CCl₄-hepatotoxicity in the absence of increased lethality. Normal liver response to a low-dose CCl_4 injury is not abrogated by phenobarbital + CCl_4 interaction. Instead, the early phase of cell division is postponed (from the normal 6 hr to 24 hr). Enhanced putative mechanisms, such as increased bioactivation of CCl_4 and resultant increased lipid peroxidation, are responsible for the increased infliction of Stage I injury. Because hepatocellular regeneration and tissue repair processes continue, albeit a bit later than normal, these hormetic mechanisms permit tissue restoration, resulting in recovery from the enhanced liver injury. This mechanism explains the remarkable recovery from phenobarbital-induced enhancement of CCl_4 liver injury. A remarkably enhanced liver injury by phenobarbital is of no real consequence to the animal's survival because depletion of cellular energy does not occur with this interaction, which permits hormetic mechanisms to restore hepatolobular architecture, resulting in complete recovery.

and restoration of hepatolobular architecture. The prolongation of these normal responses of the liver is a consequence of the enhanced liver injury, inflicted by the enhanced putative injurious mechanisms. The hypothesis that suppression of hepatocellular regeneration and tissue repair leads to the progression of liver injury was experimentally validated in a partial hepatectomy model.[81-84,86]

³H-Thymidine incorporation into nuclear DNA and the labeling index, as evidenced by autoradiography of liver sections, were significantly increased in rats maintained on normal diet at 1 to 2 hr after CCl₄ administration (100

μL/kg) and returned to basal level by 6 hr.[84] Serum transaminases in these rats undergoing sham operation and partial hepatectomy were not altered. However, these enzymes were significantly elevated in chlordecone-treated rats following CCl_4 administration (100 μL/kg). CCl_4-induced serum enzyme elevations were significantly lower in two-days post–partial hepatectomy rats than in sham operated rats. If the early-phase (6-hr) stimulation of hepatocellular regeneration is critical for recovery, protection by partial hepatectomy should be phased out by seven days, since stimulated hepatocellular regeneration phases out by then.[81] Seven-days post–partial hepatectomy rats maintained on chlordecone diet were not protected from chlordecone-potentiated CCl_4 hepatotoxicity, indicating the importance of the early-phase tissue repair.[82,83] The protection against hepatotoxic and lethal effects of the chlordecone + CCl_4 combination by previously stimulated hepatocellular regeneration might be explained by two consequences of stimulated cell division:

1. The hepatocellular architecture is renovated by the newly divided cells.
2. Because of the well-known resistance of the newly divided cells,[39,116-118] the permissive progression of toxicity is obtunded.

The pivotal importance of the early-phase stimulation of hepatocellular regeneration is also evident when one considers that a large (or massive) dose of CCl_4 is toxic because of the suppression of the early phase tissue repair, since the second wave of hepatocellular regeneration (48 hr) remains intact.[84]

Additional Evidence in Support of a Critical Role for the Early-Phase Stimulation of Cell Division and Tissue Repair

Table 4.5 presents a variety of experimental manipulations that permit a rigorous experimental verification of the existence of, and the critical role played by, inducible tissue repair in the final outcome of toxic injury. All of the evidence for the existence of a hormetic mechanism was derived through efforts to understand the mechanism of chlordecone potentiation of halomethane toxicity (Table 4.5).

Partial Hepatectomy

If the basic premise, that suppression of the early-phase (6-hr) stimulation of cell division and tissue repair is the mechanism of chlordecone potentiation of CCl_4 injury, is valid, then a preplacement of cell division in the liver should result in protection against the interactive toxicity of chlordecone + CCl_4. When CCl_4 was administered at 2 days after partial hepatectomy at a time of maximally stimulated hepatocellular division, a remarkable protection was observed.[81] At 7 days after partial hepatectomy, when the stimulated cell division phases out, the interactive toxicity becomes fully manifested again.[81] In these studies, microsomal cytochrome

Table 4.5. Evidence in Support of the Proposed Mechanism

Experimental Manipulation	Findings	References
1. Preplaced cell division and tissue repair by partial hepatectomy	Protection from chlordecone + CCl_4	81–84
2. Toxicity of a large dose of CCl_4	Early-phase stimulation of tissue repair is ablated	84,88
3. Hepatocytes isolated from chlordecone-treated rats incubated with CCl_4 (isolated hepatocytes do not divide in vitro)	No potentiation in contrast to in vivo	119
4. Developing young rats have growing livers	Resilient to chlordecone + CCl_4	121
5. Gerbils lack the early-phase tissue repair	Low dose of CCl_4 is highly toxic	41,42
Do not have early-phase tissue repair to suppress	Chlordecone does not potentiate CCl_4 toxicity	41,42
Preplaced tissue repair by partial hepatectomy	Resiliency to CCl_4 toxicity	122
6. CCl_4 autoprotection	Due to prestimulation of early-phase tissue repair by the protective dose	48
7. Selective ablation of the early-phase hormesis by colchicine	Prolongation of hepatotoxicity of a low dose of CCl_4 by 24 hr (until the second phase of cell division at 48 hr ensues to overcome injury)	128–130
Colchicine given 2 hr before the protective dose of CCl_4	Abolishes CCl_4 autoprotection entirely	131

P-450 content is decreased by partial hepatectomy, but remains at the decreased level even 7 days later, when protection is no longer evident. Moreover, actual in vivo bioactivation, and overall disposition of $^{14}CCl_4$, are unperturbed by partial hepatectomy.[86]

Toxicity of Large Dose is Due to Ablation of the Hormetic Response

An implication of these findings is that the toxic effect of a large dose of CCl_4 might be a consequence of suppressed early-phase cell division and tissue repair. When a large dose of CCl_4 was administered, the early-phase cell division[81-85,108,109] normally stimulated by a low dose of CCl_4 was ablated entirely.[48,84,88] These findings indicate that the real difference between a low and a high dose of CCl_4 is the presence or absence of hormetic response in the form of stimulated early-phase cell division and tissue repair. The higher dose clearly prevents the hormetic response, thus permissively allowing toxicity to progress unabatedly, much like the unabated progression of a brushfire to a forest fire in the absence of fire fighters.

Interactive Toxicity of Chlordecone + CCl₄ Not Occurring under In Vitro Conditions Where Tissue Hormesis Cannot Be Expressed

Yet another line of experimental validation of the critical role of suppressed cell division and tissue repair comes from in vitro incubation of hepatocytes isolated from chlordecone-pretreated rats with CCl_4.[119] Isolated hepatocytes do not divide under in vitro conditions. Therefore, if suppression of cell division and tissue repair ordinarily stimulated by a low dose of CCl_4 is the mechanism of chlordecone-amplified CCl_4 toxicity, one should not observe highly amplified toxicity when hepatocytes from chlordecone-treated rats are incubated with CCl_4 in vitro. Since prior exposure to phenobarbital is known to result in increased CCl_4 toxicity in vitro, incubation of hepatocytes obtained from phenobarbital-treated rats with CCl_4 should result in a measurable level of increased toxicity. Recent experiments revealed no significant increase in cytotoxic injury in chlordecone-pretreated isolated hepatocyte incubations.[119] Cells from phenobarbital-pretreated rats exhibited highest CCl_4 toxicity, indicating that the in vitro paradigm was working as expected. These findings are consistent with the hypothesis that suppression of hepatocellular division and tissue repair is the mechanism of chlordecone-potentiated CCl_4 toxicity, and provide substantial evidence against any significant role for chlordecone-enhanced bioactivation of CCl_4.[119]

Resiliency of Newborn and Developing Rats

Newborn and young, developing rats have actively growing livers. Since livers during active growth would be expected to have ongoing cell division, these developing rats would be expected to be resilient to chlordecone potentiation of CCl_4 injury during their early development. When rat pups at 2, 5, 20, 35, 45, and 60 days were tested, rats were completely resilient to chlordecone potentiation of CCl_4 toxicity up to 35 days of age.[121] At 45 days, young rats were sensitive to the interactive toxicity of chlordecone + CCl_4, and by 60 days the rats were just as sensitive as adults.[122] The hepatic microsomal cytochrome P-450 levels in the livers of 35-, 45-, and 60-day-old rats exposed to chlordecone were not different from each other, suggesting that any differences in cytochrome P-450 levels are unlikely to explain the observed differences in toxicities. Moreover, recent studies indicate that bioactivation of $^{14}CCl_4$ in 35-day-old rats is not less than that observed in 60-day-old rats (unpublished data). Therefore, the resiliency of younger rats to chlordecone potentiation of CCl_4 toxicity is more likely related to the

Table 4.6. High Sensitivity of Mongolian Gerbils to Halomethane Toxicity Contrasted with Their Resiliency to Potentiation by Exposure to Other Chemicals

	15-Day Dietary Pretreatment (μL/kg)[a]			
Halomethane	Normal Diet	Chlordecone (10 ppm)	Phenobarbital (225 ppm)	Mirex (10 ppm)
CCl_4	80 (34–186)	100 (78–128)	100 (28–354)	100 (28–354)
$CBrCl_3$	20 (8.6–46.5)	20 (16.4–24.4)	20 (10.4–38.4)	16.8 (9.9–28.6)
$CHCl_3$	400 (208–769)	565 (346–923)	400 (268–597)	400 (268–597)

Source: Adapted from Cai and Mehendale.[42]
[a]Numbers in parentheses are 95% confidence intervals.

ongoing hepatocellular regeneration during early development rather than due to differences in the bioactivation of CCl_4.

Gerbils' Lack of the Early-Phase Hormesis and Greatest Sensitivity to Halomethane Toxicity

While administration of a low dose of CCl_4 to rats results in a prompt stimulation of early-phase hepatocellular regeneration at 6 hr,[81-85] in Mongolian gerbils this early-phase cell division is not observed.[42] The stimulation of cell division that does occur at 42 hr (analogous to the second phase of cell division, which occurs at 48 hr in rats) appears to be too little and too late to be of any help in overcoming liver injury.[41,42] If the early-phase cell division is critical for recovery from liver injury, then because of their lack of this important hormetic mechanism, gerbils would be expected to be extremely sensitive to halomethane toxicity. When tested, gerbils were found to be approximately 35-fold more sensitive to the toxicity of CCl_4, $BrCCl_3$, and $CHCl_3$ (Table 4.6).[41] It follows that gerbils should not be susceptible to chlordecone potentiation of CCl_4 toxicity (Table 4.6) since they lack the early phase of hepatocellular regeneration.[42]

Subsequent studies have shown that a preplacement of hepatocellular regeneration by partial hepatectomy results in significant protection against CCl_4 toxicity,[122] underscoring the importance of stimulated hepatocellular regeneration in determining the final outcome of liver injury. These studies also reveal another important difference between species. While rats respond by maximal stimulation of hepatocellular regeneration within 2 days after partial hepatectomy, in gerbils the maximal stimulation was many fold lower and it occurred not before 5 days after partial hepatectomy.[122] These findings indicate that gerbils are much more sluggish in their hormetic response to a noxious challenge of a hepatotoxic chemical agent. Each of these findings (Table 4.5) is consistent with the critical importance of the early-phase stimulation of cell division as a decisive target of inhibition in chlordecone potentiation of CCl_4 toxicity. Further, these findings also underscore the importance of the biological hormetic response in determining the resiliency to the toxic action of halomethanes.

Autoprotection

CCl_4 autoprotection is a phenomenon whereby administration of a single low dose of CCl_4 24 hr prior to the administration of a killing dose of the same compound results in an abolition of the killing effect of the large dose.[43-45,123,124] The widely accepted mechanism of this phenomenon is the destruction of liver microsomal cytochrome P-450 by the protective dose such that the subsequently administered large dose is insufficiently bioactivated.[12,13,19,46,47,49,50] Since bioactivation of CCl_4 is an obligatory step for its necrogenic action, it was suggested that massive liver injury ordinarily expected from a large dose of CCl_4 never occurs in the autoprotected animal.[12] Although this mechanism has been widely accepted, a closer examination of the evidence suggests that the mechanism was largely derived by association[12,13,43-46,123,124] rather than actual experimental evidence of less-than-expected liver injury in the autoprotected animal.

Additionally, several lines of evidence indicate that even after the significant destruction of cytochrome P-450, the availability of the P-450 isozyme responsible for the bioactivation of CCl_4 is not limiting.[48,81,86,122,125,126] For instance, even after a 60% decrease in the constitutive liver microsomal cytochrome P-450 by $CoCl_2$ treatment, CCl_4 toxicity was undiminished regardless of whether the rats were pretreated with chlordecone.[81] More direct evidence was obtained from studies in which in vivo metabolism and bioactivation of $^{14}CCl_4$ was examined in rats pretreated with $CoCl_2$.[86] The uptake, metabolism, and bioactivation of CCl_4 was not significantly altered in $CoCl_2$-treated rats known to have highly decreased liver microsomal cytochrome P-450 content.

Additional experimental evidence, indicating that actual liver injury observed in rats receiving a high dose of CCl_4 was identical regardless of whether a prior protective dose was administered, led to a reexamination of the mechanism underlying CCl_4 autoprotection.[48] A systematic time-course study, in which biochemical and histopathological parameters, as well as animal survival, were examined, revealed a critical role for the hormetic response of the liver in the form of stimulated early-phase cell division and tissue repair.[48] The protective dose-stimulated tissue repair results in augmented and sustained hepatocellular regeneration and tissue repair, which enable the autoprotected rats to overcome the same level of massive injury, which is ordinarily irreversible and leads to hepatic failure followed by animal death.[48,125,127]

*Selective Ablation of the Early-Phase Hormetic Response
by Colchicine*

Finally, the pivotal importance of the early-phase stimulation of hepato-cellular division and tissue repair can be tested with an elegant experimental tool, colchicine. With a carefully selected dose of colchicine, it is possible to selectively ablate the early-phase stimulation of mitosis associated with the administration of a low dose of CCl_4.[128] One single administration of colchi-cine at 1 mg/kg results in ablation of mitotic activity — the effect lasting only up to 12 hr — such that the second phase of cell division at 48 hr after the administration of CCl_4 remains unperturbed.[129] At this dose colchicine does not cause any detectable liver injury nor does it cause any adverse perturbation of hepatobiliary function.[130] Therefore, under these condi-tions, use of colchicine permits a very important experimental paradigm in which the early-phase hormesis in response to a low dose of CCl_4 can be selectively ablated.

Using the model of colchicine antimitosis, the importance of the early-phase stimulation of hepatocellular regeneration in the toxicology of CCl_4 was tested.[128] The selective ablation of the early-phase response of cell division resulted in a prolongation of limited liver injury associated with a low dose of CCl_4.[129] Ordinarily, intraperitoneal administration of 100 μL CCl_4/kg results in very limited liver injury, which is overcome by stimulated cell division and tissue repair within 24 hr.[81-85,108,109] The prolongation of this limited injury lasted only for an additional 24 hr (up to 48 hr after CCl_4 injection), at which time the unperturbed second phase of cell division permits complete recovery to occur within the next 24 hr (by 72 hr after CCl_4 injection). This increased and prolonged CCl_4 injury is not accompa-nied by enhanced bioactivation of CCl_4.[128,129] Indeed, actual liver injury assessed by morphometric analysis of hepatocellular necrosis and ballooned cells is not enhanced during the first 12 hr in colchicine-treated rats, further indicating that enhancement of the mechanisms responsible for infliction of injury was not involved.[128-130] These findings underscore the pivotal role of the early-phase stimulation of hormesis in the final outcome of toxicity associated with a low dose of CCl_4.

Another experimental paradigm permits a further test of how critical the early-phase hormetic response is in the final outcome of injury. In the above-described experiments, the preservation of the second phase of cell division permits complete recovery by 72 hr. Administration of a large dose of CCl_4 permits experimental interference with this second phase of cell division. In such an experiment, the animals should not survive because of continued progression of toxicity; in other words, selective ablation of the early-phase hormetic response in an autoprotection protocol should result in a denial of autoprotection. Indeed, 100% survival observed in an experi-mental protocol (100 μL CCl_4/kg administered 24 hr prior to the injection of 2.5 mL CCl_4/kg) is completely denied by colchicine antimitosis.[131] This observation also provides very substantial and convincing experimental evidence for the newly proposed mechanism for the autoprotection phenomenon.[48,127] The mechanism underlying the autoprotection phenomenon is the ability of the liver tissue to respond by augmentation of tissue repair through hormesis induced by the protective dose.[48]

A Two-Stage Model of Toxicity

An intriguing outcome of the work on the interactive toxicity of chlordecone + CCl_4 is the emergence of a concept that permits the separation of the early events responsible for initiating injury from subsequent events that determine the final outcome of that injury (Figure 4.6). Hormetic mechanisms[7] are activated upon exposure to low levels of halomethanes.[82,83,85,103,108,109,132-134] Although the cellular mechanisms responsible for triggering a dramatic mobilization of biochemical events leading to cellular proliferation within 6 hr after exposure to a subtoxic dose of CCl_4 are not understood,[82,83,85,103,110] it is clear that these early events are the critical determinants of the final outcome of injury.[2,14-16] When this early phase of hepatocellular division is suppressed, as has been observed in animals pretreated with chlordecone,[82,83,85,108,109] a permissive and unabated progression of liver

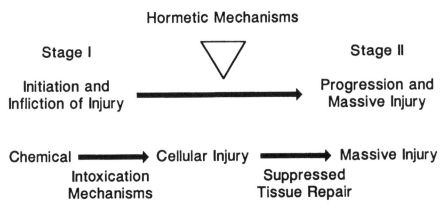

Figure 4.6.   Scheme illustrating the proposed two-stage model of toxicity. Stage I involves infliction of cellular and/or tissue injury by intoxication mechanisms, which are understood for many chemical and physical agents. If Stage I injury is inflicted by a low dose of the offending agent, hormetic mechanisms are stimulated (such as cellular regeneration and tissue repair targeted for restoration of tissue structure), and complete recovery from injury follows with no additional toxic consequence. If hormetic mechanisms are suppressed or ablated, the limited injury associated with exposure to a low dose of the offending toxic agent would continue unabated, resulting in progressive injury. High doses of toxic agents can cause ablation of the hormetic mechanism, as in the case of a high dose of CCl_4, which results in ablation of the early-phase hormetic response.[84] Another example is the ablation of the early-phase hormesis exemplified by the interactive toxicity of chlordecone and the halomethane solvents. Adapted from Mehendale.[2]

injury, leading to massive coagulative hepatic necrosis, is observed.[2,14-16] Likewise, it has been demonstrated experimentally that restoring the tissue hormesis (Figure 4.7) results in an obtundation of the progressive phase of injury, permitting the tissue to overcome injury.

The central role of hormetic mechanisms in the final outcome of tissue injury becomes self-evident from the following lines of experimental evidence. Prior exposure to 225 ppm phenobarbital results in the potentiation of liver injury by the same subtoxic dose of CCl_4 employed in the chlordecone + CCl_4 interaction.[16,55,56,85] The quantitative measures of liver injury at 24 hr after the administration of CCl_4 indicate that the tissue injury is either equivalent to or slightly greater than that seen in chlordecone + CCl_4 interaction.[55] Left alone, the animals undergoing the interactive toxicity of phenobarbital + CCl_4 recover, while those experiencing the chlordecone + CCl_4 interaction do not.[2,14-16,56] While the enhanced liver injury observed with the interactive toxicity of phenobarbital + CCl_4 is consistent with the increased bioactivation of CCl_4,[55,106] recovery from this injury is consistent

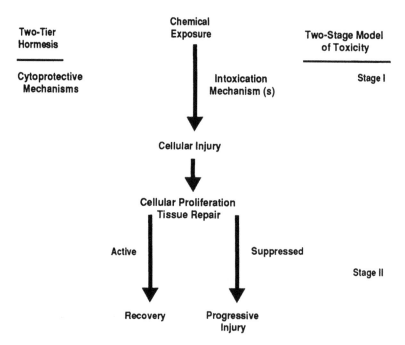

Figure 4.7. Scheme illustrating the concept of separating those mechanisms that are responsible for the infliction of cellular and tissue injury from those that follow these events. Intoxication mechanisms result in infliction of injury during Stage I of toxicity. During Stage II of toxicity, tissue hormetic mechanisms are stimulated in an attempt to overcome injury. If these hormetic mechanisms are unperturbed, recovery occurs. Interference with these mechanisms results in uncontrollable progression of injury, much like an unquenched brushfire progressing to become a forest fire.

with the unablated hepatocellular proliferation and tissue repair.[83,85] Delayed hepatocellular regeneration and tissue repair from the normal 6 hr to 24–36 hr is the only consequence on Stage II of CCl_4 toxicity.[16,85] Nevertheless, the highly stimulated early phase of tissue repair, even though it is postponed by 24 hr, enables the restoration of hepatolobular structure and function,[2,14-16,85] and thereby animal survival. These observations provide additional support for the presence of two distinct stages of chemical toxicity (Figure 4.7).

Induction of liver regeneration 36 to 48 hr after the administration of a toxic dose of CCl_4 is well established.[135-137] The existence of an early phase of cell division (6 hr) was revealed only through experiments with a low, subtoxic dose of CCl_4.[82,83,108,109] Administration of a large, toxic dose of CCl_4 (2.5 mL/kg) results in complete suppression of this early phase of cell division,[48,84,88] indicating that the toxicity associated with a large dose is due to the abolishment of this critical early-phase stimulation of tissue repair.[2,14-16] Thus, experimentally, it is possible to ablate the early phase of hepatocellular regeneration and tissue repair ordinarily stimulated by a low dose of CCl_4 — making it, in essence, a toxic dose. Administration of the same dose to animals prestimulated by partial hepatectomy, so that they have the ongoing hepatocellular proliferation and tissue repair, results in a remarkable and substantial protection from liver injury and lethality.[84] Such protection is not due to decreased bioactivation of CCl_4.[86]

The importance of the stimulation of tissue repair as an event independent of Stage I of chemical toxicity can be illustrated by other elegant experimental approaches. Experimental interference with the early phase of hepatocellular proliferation leads to prolonged and enhanced liver injury of an ordinarily subtoxic dose of CCl_4. Studies with colchicine antimitosis,[128-131] in which colchicine administered selectively ablate the early phase of hepatocellular division (6 hr) without interfering with the second phase of hepatocellular regeneration (48 hr), have shown a prolongation of liver injury. Neither liver injury measured through serum enzyme elevations nor by morphometric analysis of necrosis was increased at 6 or 12 hr in colchicine-treated rats — findings consistent with the lack of colchicine-enhanced bioactivation of CCl_4.[129,131] Moreover, colchicine ablation of the early-phase hormetic response after the protective dose of CCl_4 in an autoprotection protocol leads to complete denial of autoprotection.

The critical role played by the capacity to respond to CCl_4 hepatotoxicity by stimulation of tissue repair mechanisms at an early time point is illustrated by examining species and strain differences in susceptibility to CCl_4 injury. Mongolian gerbils are extremely sensitive to halomethane hepatotoxicity.[40-42,122,138] Gerbils are approximately 35-fold more sensitive to CCl_4 toxicity than Sprague-Dawley rats.[41,42] This difference in CCl_4 toxicity can be seemingly explained on the basis of a 3.5-fold greater bioactivation of CCl_4 in gerbils.[41] However, the remarkable and substantial sensitivity does not appear to be due to 3.5-fold greater bioactivation of CCl_4, since CCl_4

toxicity is not at all increased in gerbils by prior exposure to phenobarbital in spite of a 5-fold greater bioactivation of CCl_4.[41,42] The time-course studies on the ability of gerbils to respond to a subtoxic dose of CCl_4 by stimulation of hepatocellular regeneration and tissue repair reveal an important difference in the biology of the hormetic mechanisms between gerbils and rats.[42] The early-phase stimulation of tissue repair in the liver does not manifest itself in gerbils, and the second phase occurs approximately 40 hr after the administration of CCl_4.[42,122] In the absence of the biological mechanism to arrest the progression of liver injury (Figure 4.7), the liver injury might be expected to permissively progress much like an unquelched brushfire.

Evidence in support of the concept that species differences in chemical toxicity might depend on the differences in the promptness in initiating tissue repair mechanisms among various species comes from another aspect of the interactive toxicity of chlordecone + CCl_4. While gerbils are extremely sensitive to CCl_4, this sensitivity cannot be further increased by prior exposure to chlordecone.[41,42,122,138] Since substantial evidence supports the concept that suppression of the early phase of hepatocellular regeneration and tissue repair is the mechanism for the permissive progression of liver injury in the chlordecone + CCl_4 interaction,[2,14-16] lack of this early-phase response in the gerbil would be consistent with extremely high sensitivity of gerbils to CCl_4 on the one hand, and a lack of potentiation of CCl_4 toxicity by prior exposure to chlordecone on the other.[41,42] This concept has received additional support recently through partial hepatectomy experiments.[122]

The interactive toxicity of chlordecone + $CHCl_3$ has been demonstrated in murine species.[100-103] Stimulation of hepatocellular regeneration and tissue repair after a subtoxic dose of $CHCl_3$ allows the mice to overcome the liver injury associated with that dose of $CHCl_3$.[103] By lowering the dose of $CHCl_3$ used in the chlordecone + $CHCl_3$ studies,[102] it is possible to demonstrate potentiation of liver injury, but without the lethality.[103] Such an experimental protocol vividly reveals a decisive role played by the stimulated tissue repair mechanisms in overcoming liver injury[103] and the separation of these mechanisms (Stage II) from the inflictive phase (Stage I) of chemical injury (Figure 4.7).

The importance of stimulated tissue repair mechanisms in overcoming liver injury has also been demonstrated through examination of the mechanistic basis for significant strain differences in mice.[139,140] A SJL/J strain of mice, known to be less susceptible to CCl_4 toxicity, was shown to possess more prompt and efficient tissue repair mechanisms, which permit augmented recovery, while the BALB/C strain, known to be more susceptible, was shown to possess less efficient tissue repair mechanisms, resulting in retarded recovery.[139] The F_1 cross between these two strains was shown to be intermediate in susceptibility.[140] A careful histopathological evaluation revealed that while the time course of the appearance in injury was quite similar (Stage I, Figure 4.6), significant differences in tissue repair mechanisms between these strains could account for the strain differences in the ultimate toxicity of CCl_4.[139,140] While the time course of the inflictive phase of injury in the F_1 (SJL/J × BALB/C) was similar to the two parent strains, the tissue repair was at the intermediate level of augmented (SJL/J) and retarded (BALB/C) recovery.

With the advent of the finding that a low dose of CCl_4 is not toxic—not so much because it does not initiate tissue injury, but because of the stimulated tissue repair mechanisms[108,109]—it became apparent that the stimulation of the early phase of hepatocellular regeneration is, in essence, an endogenous hormetic mechanism, recruited to overcome tissue injury. One implication of this finding is its possible role in the phenomenon of CCl_4 autoprotection.[43-46] Circumstantial evidence, in which a decrease in hepatic microsomal cytochrome P-450 by $CoCl_2$ administration to 40% of the normal level did not result in decreased CCl_4 liver injury,[81] suggested the possibility that mechanism(s) other than decreased cytochrome P-450 might be involved in CCl_4 autoprotection. Recent studies reveal a critical role for the hepatocellular regeneration and tissue repair stimulated by the low protective dose administration.[48] Essentially, the protective dose serves to stimulate tissue repair mechanisms[82,83,86,108,109] so that even before the large dose known to abolish the early-phase stimulation of tissue repair[84] is administered, the tissue repair mechanisms are already in place, resulting in augmentation of tissue repair sufficient to tip the balance between progression of injury and recovery in favor of the latter.[48] This experimental model represents another example in which a selective augmentation of the tissue hormetic mechanism (Stage II, Figure 4.6), independent of the inflictive phase of toxicity (Stage I, Figure 4.6), can dramatically alter the ultimate outcome of toxic injury (Figure 4.7).

Another line of evidence to implicate the importance of the hormetic mechanisms in determining the final outcome of chemical toxicity comes from experiments designed to understand the mechanisms responsible for the failure of the tissue regenerative and repair mechanism in the interactive toxicity of chlordecone + CCl_4. Much evidence is available pointing to insufficient availability of cellular energy at a time when cell division should have taken place.[87,108,109] A remarkable and irreversibly precipitous decline in glycogen levels in the liver,[109,114] a rise in hepatocellular Ca^{2+},[110-113] and a consequent stimulation of phosphorylase a activity, leading to an equally precipitous decline in hepatic ATP[114,115] are events consistent with the failure of hepatocellular regeneration in the chlordecone + CCl_4 interaction. Only marginal and transient declines in ATP levels in the interactive hepatotoxicity of phenobarbital + CCl_4 and mirex + CCl_4 are consistent with a postponement of hepatocellular regeneration,[115] leading to transiently increased liver injury followed by complete recovery.[85] The concept of insufficient hepatocellular energy being linked to failure of hepatocellular regeneration and tissue repair has gained support from experiments in which the administration of an external source of energy resulted in augmented ATP levels

Table 4.7. Chemicals Reported to Cause Nonneoplastic Hepatocellular Proliferation

Chemicals	References
1. Acetaminophen	3
2. Allyl alcohol	3,4
3. a-Naphthyl isothiocyanate	141,142
4. Bromotrichloromethane	132
5. Carbon tetrachloride	108,109,137
6. Chloroform	143
7. Ethylene dibromide	144
8. Galactosamine	145,146
9. Thioacetamide	147,148

and significant protection.[88-90] In recent studies, catechin (cyanidanol), known to increase hepatic ATP levels, provided substantial protection against the lethal effect of chlordecone + CCl_4.[89,90] Protection by catechin was accompanied by a restoration of the stimulation of hepatolobular repair and tissue healing.[90] The most interesting aspect of catechin protection against the interactive toxicity of chlordecone + CCl_4 is that protection does not appear to be the result of decreased infliction of hepatic injury,[89,90] as evidenced by a lack of difference in injury up to 24 hr after CCl_4 administration.[90] Cyanidanol protection was clearly due to restored hepatocellular regeneration and tissue repair. These observations provide substantial evidence for the separation of Stage I of toxicity, responsible for the infliction of tissue injury, from the Stage II events, responsible for the final outcome of tissue injury.[2]

Abundant opportunities to test the two-stage model of toxicity are available. Many chemicals have been reported to induce hepatocellular regeneration at relatively modest doses, some of which are listed in Table 4.7.[3,4,108,109,132,137,141-148] Opportunities to test the conceptual framework being put forth here are available through additional investigations with these models of tissue injury, as well as scores of other models in other tissues and organs.

Implications for Assessment of Risk to Public Health

Establishing that the initial toxic or injurious events, regardless of how they are caused, can be separated from the subsequent events that determine the ultimate outcome of injury, offers promising opportunities for developing new avenues for therapeutic intervention, with the aim of restoring the hormetic tissue repair mechanisms. Such a development will open up avenues for two types of measures to protect public health:

1. By and large, the presently used principle is to decrease the injury by interfering with Stage I of toxicity.
2. Tissue repair and healing mechanisms could be enhanced not only to obtund the progression of injury, but also to simultaneously augment recovery from that injury.

In addition to these opportunities, the two-stage concept of chemical toxicity also embodies implications of significant interest in the assessment of risk from exposure to toxic chemicals. The existence of a threshold for chemical toxicity is evident, as indicated by observation of the stimulation of tissue repair mechanisms, directed to tissue healing and recovery, after the administration of subtoxic levels of toxic chemicals, when exposure involves singular chemicals. The existence of a two-level or two-stage threshold is apparent from the two-tier hormetic response: one threshold level for each stage of the two-stage model (Figure 4.7). Generally speaking, the threshold for Stage I of toxicity must lie in the cytoprotective mechanisms (cellular hormesis). The threshold for Stage II of toxicity appears to be in the tissue's ability to respond promptly by augmenting tissue-healing mechanisms. These two thresholds may be quantitatively the same or different, but clearly have mechanistic basis at different levels.

From a public health perspective, exposure to singular chemicals is seldom involved. Multiple exposures to chemical combinations and/or singular components simultaneously, intermittently, or sequentially are almost always the rule. In this regard, antagonistic interactive toxicity or inconsequential interactions are also of interest. Of greater interest from a public health perspective is the finding that the hormetic mechanisms that constitute the threshold for physical or chemical toxicity can be mitigated in the interactive toxicology of chemical and physical agents, resulting in highly accentuated toxicity.

Of significantly greater interest to regulatory toxicology is the need to take into account in the risk assessment process the hormetic mechanisms operating particularly at the low levels of exposure to chemicals. The recognition of the existence of cellular and tissue hormesis provides the mechanistic basis to recognize thresholds for toxic effects, thereby permitting us to take into consideration the lack of recognizable adverse health effects at low levels of exposure to chemicals in our environment.

ACKNOWLEDGMENTS

The author's research was supported by grant support from the Department of the Air Force AFOSR-88–0009, the Harry G. Armstrong Aerospace Medical Research Laboratory through U.S. EPA CR-814053, and by the 1988 Burroughs Wellcome Toxicology Scholar Award.

REFERENCES

1. Klaassen, C. D., and J. Doull. "Evaluation of Safety: Toxicological Evaluation," in *Toxicology: The Basic Science of Poisons,* 4th ed., J. Doull, C. D. Klaassen, and M. O. Amdur, Eds. (New York: Pergamon Press, New York, 1991), pp. 11–27.

2. Mehendale, H. M. "Role of Hepatocellular Regeneration and Hepatolobular Healing in the Final Outcome of Liver Injury: A Two-Stage Model of Toxicity," *Biochem. Pharmacol.* 42:1155–1162 (1991).

3. Zieve, L., W. R. Anderson, C. Lyftogt, and K. Draves. "Hepatic Regenerative Enzyme Activity after Pericentral and Periportal Lobular Toxic Injury," *Toxicol. Appl. Pharmacol.* 86:147–158 (1986).

4. Zieve, L., W. R. Anderson, and D. Lafontaine. "Hepatic Failure Toxins Depress Liver Regenerative Enzymes after Periportal Injury with Allyl Alcohol in the Rat," *J. Lab. Clin. Med.* 111:725–730 (1988).

5. Luckey, T. *Radiation Hormesis in Mammals* (Boca Raton, FL: CRC Press, 1991), pp. 256.

6. Boxenbaum, H., P. Neafsey, and D. Fournier. "Hormesis, Gompertz Functions, and Risk Assessment," *Drug Metab. Rev.* 19:195–229 (1988).

7. Sagan, L. "On Radiation, Paradigms, and Hormesis," *Science* 245:574, 621 (1989).

8. Utley, W. S., and H. M. Mehendale. "Pentobarbital-Induced Cytosolic Cytoprotective Mechanisms That Offset Increases in NADPH Cytochrome P-450 Reductase Activity in Menadione-Mediated Cytotoxicity," *Toxicol. Appl. Pharmacol.* 99:323–333 (1989).

9. Utley, W. S., and H. M. Mehendale. "Phenobarbital-Induced Cytoprotective Mechanisms in Menadione Metabolism: The Role of Glutathione Reductase and DT-Diaphorase," *Int. J. Biochem.* 22:957–967 (1990).

10. Utley, W. S., and H. M. Mehendale. "Cytoprotective Mechanisms That Offset Phenobarbital-Induced Increment in O_2 Generated from Quinone Recycling," in *Proceedings of the International Conference on Biological Oxidation Systems,* Vol. 1, C. C. Reddy, G. Hamilton, and K. M. Madhyastha, Eds. (San Diego, CA: Academic Press, 1990), Chapter 1, pp. 183–200.

11. Thor, H., M. T. Smith, P. Hartzell, G. Bellomo, S. A. Jewell, and S. Orrenius. "The Metabolism of Menadione (2-Methyl-1,4-naphthoquinone) by Isolated Hepatocytes," *J. Biol. Chem.* 257:12419–12425 (1982).

12. Recknagel, R. O., and E. A. Glende, Jr. "Lipid Peroxidation: A Specific Form of Cellular Injury," in *Handbook of Physiology,* D. H. K. Lee, Ed. (Bethesda, MD: American Physiological Society; Baltimore, MD: Williams and Wilkins, 1977), Section 9, pp. 591–601.

13. Slater, R. F. "Free Radicals and Tissue Injury: Fact and Fiction," *Br. J. Cancer (Suppl.)* 8:5–10 (1987).

14. Mehendale, H. M. "Amplification of Hepatotoxicity and Lethality of CCl_4 and $CHCl_3$ by Chlordecone," *Rev. Biochem. Toxicol.* 10:91–138 (1989).

15. Mehendale, H. M. "Mechanism of the Lethal Interaction of Chlordecone and CCl_4 at Nontoxic Doses," *Toxicol. Lett.* 49:215–241 (1989).

16. Mehendale, H. M. "Potentiation of Halomethane Hepatotoxicity by Chlordecone: A Hypothesis for the Mechanism," *Med. Hypoth.* 33:289–299 (1990).

17. Slater, T. F. "Necrogenic Action of Carbon Tetrachloride in the Rat: Speculative Mechanism Based on Activation," *Nature (London)* 209:36–40 (1966).

18. Koch, R. R., E. A. Glende, Jr., and R. O. Recknagel. "Hepatotoxicity of Bromotrichloromethane-Bond Dissociation Energy and Lipid Peroxidation," *Biochem. Pharmacol.* 23:2907–2915 (1974).

19. Sipes, I. G., G. Krishna, and J. R. Gillette. "Bioactivation of Carbon Tetra-

chloride, Chloroform and Bromotrichloromethane: Role of Cytochrome P-450," *Life Sci.* 20:1541-1548 (1974).

20. Connor, H. D., R. G. Thurman, M. D. Galiz, and R. P. Mason. "The Formation of a Novel Free-Radical Metabolite from CCl₄ in the Perfused Rat Liver and In Vivo," *J. Biol. Chem.* 261:4542-4548 (1986).

21. Manno, M., F. De Matteis, and L. J. King. "The Mechanism of the Suicidal, Reductive Inactivation of Microsomal Cytochrome P-450 by Carbon Tetrachloride," *Biochem. Pharmacol.* 37:1981-1990 (1988).

22. Moody, D. E., J. L. James, and E. A. Smuckler. "Phenobarbital Pretreatment Alters the Localization of CCl₄-Induced Changes in Rat Liver Microsomal Fatty Acids," *Toxicol. Appl. Pharmacol.* 103:16-17 (1990).

23. Vittozzi, L., and W. Nastainczyk. "Binding of Reactive Metabolites of CCl₄ to Specific Microsomal Proteins," *Biochem. Pharmacol.* 36:1401-1406 (1987).

24. Pohl, L. R., R. V. Branchflower, R. J. Highet, J. L. Martin, D. S. Nunn, T. J. Monks, J. W. George, and J. A. Hinson. "The Formation of Diglutathionyl Dithiocarbonate as a Metabolite of Chloroform, Bromotrichloromethane and Carbon Tetrachloride," *Drug Metab. Dispos.* 9:334-339 (1981).

25. Cameron, G. R., and W. A. E. Karunaratne. "Carbon Tetrachloride Cirrhosis in Relation to Liver Regeneration," *J. Parasitol. Bacteriol.* 42:1-22 (1936).

26. Paul, B. B., and D. Rubenstein. "Metabolism of Carbon Tetrachloride and Chloroform by the Rat," *J. Pharmacol. Exp. Ther.* 141:141-148 (1963).

27. Reynolds, E. A. "Metabolism of CCl₄-C¹⁴ in Livers of CCl₄ Poisoned Rats," *Fed. Proc.* 22:360-370 (1963).

28. Shah, H., S. P. Hartman, and S. Weinhouse. "Formation of Carbonyl Chloride in Carbon Tetrachloride Metabolism by Rat Liver In Vitro," *Cancer Res.* 39:3942-3947 (1979).

29. Bini, A., G. Vecchi, G. Vivoli, V. Vannini, and C. Cessi. "Detection of Early Metabolites in Rat Liver after Administration of CCl₄ and BrCCl₃," *Pharmacol. Res. Commun.* 7:143-149 (1975).

30. Torrielli, M., V. G. Ugazio, L. Gabriel, and E. Burdino. "Effect of Drug Pretreatment on BrCCl₃-Induced Liver Injury," *Toxicology* 2:321-326 (1974).

31. Davies, M. J., and T. F. Slater. "Electron Spin Resonance Spin Trapping Studies on the Photolytic Generation of Halocarbon Radicals," *Chem. Biol. Interact.* 58:137-147 (1986).

32. Pohl, L. R. "Biochemical Toxicology of Chloroform," *Rev. Biochem. Toxicol.* 1:79-107 (1979).

33. Testai, E., F. Gramenzi, S. Di Marzio, and L. Vittozzi. "Oxidative and Reductive Biotransformation of Chloroform in Mouse Liver Microsomes," *Arch. Toxicol. (Suppl.)* 11:42-44 (1987).

34. Ilett, K. F., W. D. Reid, I. G. Sipes, and G. Krishna. "Chloroform Toxicity in Mice: Correlation of Renal and Hepatic Necrosis with Covalent Binding of Metabolites to Tissue Macromolecules," *Exp. Mol. Pathol.* 19:215-229 (1973).

35. Bell, R. J., J. M. Brown, F. A. Meirhenry, T. A. Jorgenson, M. Robinson, and J. A. Stober. "Enhancement of the Hepatotoxicity of Chloroform in BGC3F1 Mice by Corn Oil: Implications for Chloroform Carcinogenesis," *Environ. Health Persp.* 69:49-58 (1986).

36. Docks, E. L., and G. A. Krishna. "The Role of Glutathione in Chloroform-Induced Hepatotoxicity," *Exp. Mol. Pathol.* 24:13-22 (1976).

37. Cohen, P. J., and B. Chance. "Chemiluminescence: An Index of Hepatic Lipoperoxidation Accompanying Chloroform Anesthesia," *Biochim. Biophys. Acta* 884:517–519 (1986).

38. Ekstrom, T., A. Stahl, K. Sigvardsson, and J. Hogberg. "Lipid Peroxidation In Vivo Monitored As Ethane Exhalation and Malondialdehyde Excretion in Urine after Oral Administration of Chloroform," *Acta Pharmacol. Toxicol.* 58:289–296 (1986).

39. Ruch, R. J., J. E. Klaunig, N. E. Schultz, A. B. Askari, D. A. Lacher, M. A. Pereira, and P. J. Goldblatt. "Mechanisms of Chloroform and Carbon Tetrachloride Toxicity in Primary Cultured Mouse Hepatocytes," *Environ. Health Perspect.* 69:301–305 (1986).

40. Ebel, R. E., R. L. Barlow, and E. A. McGrath. "Chloroform Hepatotoxicity in the Mongolian Gerbil," *Fundam. Appl. Toxicol.* 8:207–216 (1987).

41. Cai, Z., and H. M. Mehendale. "Lethal Effects of CCl_4 and Its Metabolism by Gerbils Pretreated with Chlordecone, Phenobarbital and Mirex," *Toxicol. Appl. Pharmacol.* 104:511–520 (1990).

42. Cai, Z., and H. M. Mehendale. "Hepatotoxicity and Lethality of Halomethanes in Mongolian Gerbils Pretreated with Chlordecone, Phenobarbital or Mirex," *Arch. Toxicol.* 65:204–212 (1991).

43. Glende, E. A., Jr. "Carbon Tetrachloride-Induced Protection against Carbon Tetrachloride Toxicity: The Role of the Liver Microsomal Drug-Metabolizing System," *Biochem. Pharmacol.* 21:1697–1702 (1972).

44. Dambrauskas, T., and H. H. Cornish. "Effect of Pretreatment of Rats with Carbon Tetrachloride in Tolerance Development," *Toxicol. Appl. Pharmacol.* 17:83–97 (1970).

45. Ugazio, G., R. R. Koch, and R. O. Recknagel. "Mechanism of Protection against Carbon Tetrachloride by Prior Carbon Tetrachloride Administration," *Exp. Mol. Pathol.* 16:281–285 (1973).

46. Lindstrom, L. D., and M. W. Anders. "Studies on the Mechanism of Carbon Tetrachloride Autoprotection: Effect of Protective Dose of Carbon Tetrachloride on Lipid Peroxidation and Glutathione Peroxidase, Glutathione-Reductase," *Toxicol. Lett.* 1:109–114 (1977).

47. Noguchi, T., K. L. Fong, E. K. Lai, S. S. Alexander, M. M. King, L. Olson, J. L. Poyer, and P. B. McCay. "Specificity of a Phenobarbital-Induced Cytochrome P-450 for Metabolism of Carbon Tetrachloride to the Trichloromethyl Radical," *Biochem. Pharmacol.* 31:615–624.

48. Thakore, K. N., and H. M. Mehendale. "Role of Hepatocellular Regeneration in Carbon Tetrachloride Autoprotection," *Toxicol. Pathol.* 19:47–58 (1991).

49. Tomasi, A., E. Albano, K. A. K. Lott, and T. F. Slater. "Spin Trapping of Free Radical Products of CCl_4 Activation Using Pulse Radiolysis and High Energy Radiation Procedures," *FEBS Lett.* 122:303–306 (1980).

50. Rosen, G. M., and E. J. Rauckman. "Carbon Tetrachloride–Induced Lipid Peroxidation: A Spin Trapping Study," *Toxicol. Lett.* 10:337–344 (1982).

51. Green, J. "Vitamin E and the Biological Antioxidant Theory," *Ann. NY Acad. Sci.* 203:29–44 (1972).

52. Benedetti, A., A. F. Casini, M. Ferrali, and M. Comporti. "Extraction and Partial Characterization of Dialyzable Products Originating from the Peroxidation of Liver Microsomal Lipids and Inhibiting Microsomal Glucose-6-phosphatase Activity," *Biochem. Pharmacol.* 28:2909–2918 (1979).

53. Ferrali, M., R. Fulceri, A. Benedetti, and M. Comporti. "Effect of Carbonyl Compounds (4-Hydroxyalkenals) Originating from the Peroxidation of Liver Microsomal Lipids and Inhibiting Microsomal Glucose-6-phosphatase Activity," *Res. Commun. Chem. Pathol. Pharmacol.* 30:99–112 (1980).

54. Harris, R. N., and M. W. Anders. "2-Propanol Treatment Induces Selectively the Metabolism of Carbon Tetrachloride to Phosgene: Implications for Carbon Tetrachloride Hepatotoxicity," *Drug Metab. Dispos.* 9:551–556 (1981).

55. Mehendale, H. M. "Potentiation of Halomethane Hepatotoxicity: Chlordecone and Carbon Tetrachloride," *Fundam. Appl. Toxicol.* 4:295–308 (1984).

56. Klingensmith, J. S., and H. M. Mehendale. "Potentiation of CCl$_4$ Lethality by Chlordecone," *Toxicol. Lett.* 11:149–154 (1982).

57. Frank, H., H. J. Haussman, and H. Remmer. "Metabolic Activation of Carbon Tetrachloride: Induction of Cytochrome P0459 with Phenobarbital or 3-Methylcholanthrene and Its Covanting Binding," *Chem. Biol. Interact.* 40:193–208 (1982).

58. Cluet, J. L., M. Boisset, and C. Boudene. "Effect of Pretreatment with Cimetidine or Phenobarbital on Lipoperoxidation in Carbon Tetrachloride and Trichloroethylene-Dosed Rats," *Toxicology* 38:91–102 (1986).

59. Klaassen, C. D., and G. L. Plaa. "Relative Effects of Various Chlorinated Hydrocarbons on Liver and Kidney Function in Mice," *Toxicol. Appl. Pharmacol.* 9:139–151 (1966).

60. Shibayama, Y. "Hepatotoxicity of Carbon Tetrachloride after Chronic Ethanol Consumption," *Exp. Mol. Pathol.* 49:234–242 (1988).

61. Pilon, D., M. Charbonneau, J. Brodeur, and G. L. Plaa. "Metabolites and Ketone Body Production Following Methyl n-Butyl Ketone Exposure as Possible Indices of Methyl-n-butyl Ketone Potentiation of Carbon Tetrachloride Hepatotoxicity," *Toxicol. Appl. Pharmacol.* 85:49–49 (1986).

62. Pilon, D., J. Brodeur, and G. L. Plaa. "Potentiation of CCl$_4$-Induced Liver Injury by Ketonic and Ketogenic Compounds: Role of the CCl$_4$ Dose," *Toxicol. Appl. Pharmacol.* 94:183–190 (1988).

63. McLean, A. E. M., and E. K. McLean. "The Effect of Diet and 1,1,1-Trichloro-bis(p-chlorophenyl)ethane (DDT) on Microsomal Hydroxylating Enzymes and on the Sensitivity of Rats to Carbon Tetrachloride Poisoning," *Biochem. J.* 100:564–571 (1966).

64. Carlson, G. P. "Potentiation of Carbon Tetrachloride Hepatotoxicity in Rats by Pretreatment with Polychlorinated Biphenyls," *Toxicology* 5:69–77 (1975).

65. Shertzer, H. G., F. A. Raitman, and M. W. Tabor. "Influence of Diet on the Expression of Hepatotoxicity from Carbon Tetrachloride in ICR Mice," *Drug-Nutrient. Interact.* 5:275–282 (1988).

66. Sidransky, H., E. Verney, R. N. Kurl, and T. Razavi. "Effect of Tryptophan on Toxic Cirrhosis Induced by Intermittent Carbon Tetrachloride Intoxication in the Rat," *Exp. Mol. Pathol.* 49:102–110 (1988).

67. Barrow, L., and M. S. Tanner. "The Effect of Carbon Tetrachloride on the Copper-Laden Rat Liver," *Br. J. Exp. Pathol.* 70:9–19 (1989).

68. Shibayama, Y. "Potentiation of Carbon Tetrachloride Hepatotoxicity by Hypoxia," *Br. J. Exp. Pathol.* 67:909–914 (1986).

69. Ebel, R. E. "Pyrazole Treatment of Rats Potentiates CCl$_4$-But Not CHCl$_3$-Hepatotoxicity," *Biochem. Biophys. Res. Commun.* 161:615–618 (1989).

70. Sipes, I. G., A. E. E. Sisi, W. W. Sim, S. A. Mobly, and D. L. Earnest. "Role

of Reactive Oxygen Species Secreted by Activated Kupfer Cells in the Potentiation of Carbon Tetrachloride Hepatotoxicity by Hypervitaminosis A," in *Cells of the Hepatic Synusoid,* E. Wisse, D. L. Knook, and K. Decker, Eds. (Rijcwijk, The Netherlands: Kupfer Cell Foundation, 1989), pp. 376–379.

71. Lamb, R. G., J. F. Borzelleca, L. W. Condie, and C. Gennings. "Toxic Interactions between Carbon Tetrachloride and Chloroform in Cultured Rat Hepatocytes," *Toxicol. Appl. Pharmacol.* 101:106–113 (1989).

72. Simmons, J. E., D. M. DeMarini, and E. Berman. "Lethality and Hepatotoxicity of Complex Waste Mixtures," *Environ. Res.* 46:74–85 (1988).

73. Suarez, K. A., G. P. Carlson, G. C. Fuller, and N. Gausto. "Differential Acute Effects of Phenobarbital and 3-Methylcholanthrene Pretreatment on CCl₄-Induced Hepatotoxicity in Rats," *Toxicol. Appl. Pharmacol.* 23:171–177 (1972).

74. Pitchumoni, C. S., R. J. Stenger, W. S. Rosenthal, and E. A. Johnson. "Effects of 3,4-Benzopyrene Pretreatment on the Hepatotoxicity of Carbon Tetrachloride in Rats," *J. Pharmacol. Exp. Ther.* 181:227–233 (1972).

75. Ferreyra, E. C., V. M. DeFenos, A. S. Bernacchi, C. R. DeCastro, and J. A. Castro. "Treatment of Carbon Tetrachloride-Induced Liver Necrosis with Chemical Compounds," *Toxicol. Appl. Pharmacol.* 42:513–521 (1977).

76. Smuckler, E. A., and T. Hultin. "Effects of SKF-525A and Adrenalectomy on the Amino Acid Incorporation by Rat Liver Microsomes from Normal and CCl₄-Treated Rats," *Exp. Mol. Pathol.* 5:504–515 (1966).

77. Perrissoud, D., and B. Testa. "Inhibiting or Potentiating Effects of Flavonoids on Carbon Tetrachloride–Induced Toxicity in Isolated Rat Hepatocytes," *Arzneimittel-Forschung* 36:1249–1253 (1986).

78. Steinhauer, L. S., J. L. Joyave, C. P. Davidson, C. K. Born, and M. E. Hamrick. "Inhibition of Carbon Tetrachloride Induced Hepatotoxicity by Dantrolene Sodium," *Res. Commun. Chem. Pathol. Pharmacol.* 52:59–70 (1986).

79. Srivasta, S. P., K. P. Singh, A. K. Saxena, P. K. Seth, and P. K. Ray. "In Vivo Protection by Protein A of Hepatic Microsomal Mixed Function Oxidase System of CCl₄ Administered Rats," *Biochem. Pharmacol.* 36:4055–4058 (1987).

80. Bursch, W., H. S. Taper, M. P. Somer, S. Meyer, B. Putz, and R. Schulte-Hermann. "Histochemical and Biochemical Studies on the Effect of the Prostacyclin Derivative Iloprost on CCl₄-Induced Lipid Peroxidation in Rat Liver and Its Significance for Hepatoprotection," *Hepatology* 9:830–838 (1989).

81. Bell, A. N., R. A. Young, V. G. Lockard, and H. M. Mehendale. "Protection of Chlordecone-Potentiated Carbon Tetrachloride Hepatotoxicity and Lethality by Partial Hepatectomy," *Arch. Toxicol.* 61:392–405 (1988).

82. Kodavanti, P. R. S., U. M. Joshi, R. A. Young, A. N. Bell, and H. M. Mehendale. "Role of Hepatocellular Regeneration in Chlordecone-Potentiated Hepatotoxicity of Carbon Tetrachloride," *Arch. Toxicol.* 63:367–375 (1989).

83. Kodavanti, P. R. S., U. M. Joshi, V. G. Lockard, and H. M. Mehendale. "Chlordecone (Kepone)-Potentiated Carbon Tetrachloride Hepatotoxicity in Partially Hepatectomized Rats. A Histomorphometric Study," *J. Appl. Toxicol.* 9:367–375 (1989).

84. Kodavanti, P. R. S., U. M. Joshi, R. A. Young, E. F. Meydrech, and H. M.

Mehendale. "Protection of Hepatotoxic and Lethal Effects of CCl$_4$ by Partial Hepatectomy," *Toxicol. Pathol.* 17:494–506 (1989).

85. Prasada Rao, K. S., and H. M. Mehendale. "Correlation of Hepatocellular Regeneration and CCl$_4$-Induced Hepatotoxicity in Chlordecone, Mirex or Phenobarbital Pretreated Rats," FASEB J. 2:A408 (1988).

86. Young, R. A., and H. M. Mehendale. "Carbon Tetrachloride Metabolism in Partially Hepatectomized and Sham Operated Rats Preexposed to Chlordecone," *J. Biochem. Toxicol.* 4:211–219 (1989).

87. Rao, S. B., and H. M. Mehendale. "Protection from Chlordecone-Potentiated CCl$_4$ Hepatotoxicity in Rats by Fructose 1,6-Diphosphate," *Int. J. Biochem.* 21:949–954 (1989).

88. Rao, S. B., and H. M. Mehendale. "Protective Role of Fructose 1,6-Bisphosphate During CCl$_4$ Hepatotoxicity in Rats," *Biochem. J.* 262:721–725 (1989).

89. Soni, M. G., and H. M. Mehendale. "Protection from Chlordecone-Amplified Carbon Tetrachloride Toxicity by Cyanidanol. Biochemical and Histological Studies," *Toxicol. Appl. Pharmacol.* 108:46–57 (1991).

90. Soni, M. G., and H. M. Mehendale. "Protection from Chlordecone-Amplified Carbon Tetrachloride Toxicity by Cyanidanol: Regeneration Studies," *Toxicol. Appl. Pharmacol.* 108:58–66 (1991).

91. Agarwal, A. K., and H. M. Mehendale. "Potentiation of Bromotrichloromethane Hepatotoxicity and Lethality by Chlordecone Preexposure in the Rat," *Fund. Appl. Toxicol.* 2:161–167 (1982).

92. Klingensmith, J. S., and H. M. Mehendale. "Potentiation of Brominated Halomethane Hepatotoxicity by Chlordecone in the Male Rat," *Toxicol. Appl. Pharmacol.* 61:429–440 (1981).

93. Kutob, S. D., and G. L. Plaa. "The Effect of Acute Ethanol Intoxication on Chloroform-Induced Liver Damage," *J. Pharmacol. Exp. Ther.* 135:245–251 (1962).

94. Kniepert, E., and V. Gorisch. "Influence of Alcohol Pretreatment on Effects of Chloroform in Rats," *Biomed. Biochim. Acta* 47:197–203 (1988).

95. Branchflower, R. V., and L. R. Pohl. "Investigation of the Mechanism of the Potentiation of Chloroform-Induced Hepatotoxicity and Nephrotoxicity by Methyl n-Butyl Ketone," *Toxicol. Appl. Pharmacol.* 61:407–413 (1981).

96. Hewitt, L. A., C. Valiquett, and G. L. Plaa. "The Role of Biotransformation-Detoxication in Acetone-2-butanone-, and 2-Hexanone-Potentiated Chloroform-Induced Hepatotoxicity," *Can. J. Physiol. Pharmacol.* 65:2313–2318 (1987).

97. Brady, J. F., D. Li, H. Ishizaki, M. Lee, S. M. Ning, F. Xiao, and C. S. Yang. "Induction of Cytochromes P-450 JJE1 and P-450 IIB1 by Secondary Ketones and the Role of P-450 JJE1 in Chloroform Metabolism," *Toxicol. Appl. Pharmacol.* 100:342–349 (1989).

98. Curtis, L. R., W. L. Williams, and H. M. Mehendale. "Potentiation of the Hepatotoxicity of Carbon Tetrachloride Following Preexposure to Chlordecone (Kepone) in the Male Rat," *Toxicol. Appl. Pharmacol.* 51:283–293 (1979).

99. Agarwal, A. K., and H. M. Mehendale. "Potentiation of CCl$_4$ Hepatotoxicity and Lethality by Chlordecone in Female Rats," *Toxicology* 26:231–242 (1983).

100. Hewitt, W. R., H. Miyajima, M. G. Cote, and G. L. Plaa. "Acute Alteration

of Chloroform-Induced Hepato- and Nephrotoxicity by Mirex and Kepone," *Toxicol. Appl. Pharmacol.* 48:509–517 (1979).

101. Hewitt, L. A., C. Palmason, S. Masson, and G. L. Plaa. "Evidence for the Involvement of Organelles in the Mechanism of Ketone-Potentiated Chloroform-Induced Hepatotoxicity," *Liver* 10:35–48 (1990).

102. Purushotham, K. R., V. G. Lockard, and H. M. Mehendale. "Amplification of Chloroform Hepatotoxicity and Lethality by Dietary Chlordecone in Mice," *Toxicol. Pathol.* 16:27–34 (1988).

103. Mehendale, H. M., K. R. Purushotham, and V. G. Lockard. "The Time-Course of Liver Injury and ^{3}H-Thymidine Incorporation in Chlordecone-Potentiated CHCl$_3$ Hepatotoxicity," *Exp. Mol. Pathol.* 51:31–47 (1989).

104. Mehendale, H. M., and V. G. Lockard. "Effect of Chlordecone on the Hepatotoxicity of 1,1,2-Trichloroethylene and Bromobenzene," *Toxicologist* 2:37 (1982).

105. Fouse, B. L., and E. Hodgson. "Effect of Chlordecone and Mirex on the Acute Hepatotoxicity of Acetaminophen in Mice," *Gen. Pharmacol.* 18:623–630 (1987).

106. Mehendale, H. M., and J. S. Klingensmith. "In Vivo Metabolism of CCl$_4$ by Rats Pretreated with Chlordecone, Mirex or Phenobarbital," *Toxicol. Appl. Pharmacol.* 93:247–256 (1988).

107. Davis, M. E., and H. M. Mehendale. "Functional and Biochemical Correlates of Chlordecone Exposure and Its Enhancement of CCl$_4$ Hepatotoxicity," *Toxicology* 15:91–103 (1980).

108. Lockard, V. G., H. M. Mehendale, and R. M. O'Neal. "Chlordecone-Induced Potentiation of Carbon Tetrachloride Hepatotoxicity: A Light and Electron Microscopic Study," *Exp. Mol. Pathol.* 39:230–245 (1983).

109. Lockard, V. G., H. M. Mehendale, and R. M. O'Neal. "Chlordecone-Induced Potentiation of Carbon Tetrachloride Hepatotoxicity: A Morphometric and Biochemical Study," *Exp. Mol. Pathol.* 39:246–256 (1983).

110. Agarwal, A. K., and H. M. Mehendale. "CCl$_4$-Induced Alterations in Ca^{2+} Homeostasis in Chlordecone and Phenobarbital Pretreated Animals," *Life Sci.* 34:141–148 (1984).

111. Agarwal, A. K., and H. M. Mehendale. "Excessive Hepatic Accumulation of Intracellular Ca^{2+} in Chlordecone Potentiated CCl$_4$ Toxicity," *Toxicology* 30:17–24 (1984).

112. Carmines, E. L., R. A. Carchman, and J. F. Borzelleca. "Kepone: Cellular Sites of Action," *Toxicol. Appl. Pharmacol.* 49:543–550 (1979).

113. Agarwal, A. K., and H. M. Mehendale. "Effect of Chlordecone on Carbon Tetrachloride–Induced Increase in Calcium Uptake in Isolated Perfused Rat Liver," *Toxicol. Appl. Pharmacol.* 83:342–348 (1986).

114. Kodavanti, P. R. S., U. P. Kodavanti, and H. M. Mehendale. "CCl$_4$-Induced Alterations in Hepatic Calmodulin and Free Calcium Levels in Rats Pretreated with Chlordecone," *Hepatology* 13:230–238 (1991).

115. Kodavanti, P. R. S., U. P. Kodavanti, and H. M. Mehendale. "Altered Hepatic Energy Status in Chlordecone (Kepone) Potentiation of CCl$_4$ Hepatotoxicity," *Biochem. Pharmacol.* 40:859–866 (1990).

116. Roberts, E., B. M. Ahluwalia, G. Lee, C. Chan, D. S. R. Sarma, and E. Farber. "Resistance to Hepatotoxins Acquired by Hepatocytes during Liver Regeneration," *Cancer Res.* 43:28–34 (1983).

117. Chang. L. W., M. A. Pereira, and J. E. Klaunig. "Cytotoxicity of Halogenated Alkanes in Primary Cultures of Rat Hepatocytes from Normal Partial Hepatectomized and Preneoplastic/Neoplastic Liver," *Toxicol. Appl. Pharmacol.* 80:274–280 (1985).

118. Ruch, R. J., J. E. Klaunig, and M. A. Pereira. "Selective Resistance to Cytotoxic Agents in Hepatocytes Isolated from Partially Hepatectomized and Neoplastic Mouse Liver," *Cancer Lett.* 26:295 (1985).

119. Mehendale, H. M., Z. Cai, and S. D. Ray. "Paradoxical Toxicity of CCl_4 in Isolated Hepatocytes from Chlordecone, Phenobarbital and Mirex Pretreated Rats," *In Vitro Toxicol.* 4: in press (1991).

120. Mehendale, H. M., and Z. Cai. "Resiliency of Rats Developing Rats to Potentiation of Hepatotoxicity and Lethality of CCl_4 Autoprotection," *Toxicologist* 11:219 (1991).

121. Cai, Z., and H. M. Mehendale. "Role of Ongoing Versus Stimulated Hepatocellular Regeneration in Resiliency to Amplification of CCl_4 Toxicity by Chlordecone," *FASEB J.* 5:A1248 (1991).

122. Cai, Z., and H. M. Mehendale. "Prestimulation of Hepatocellular Regeneration by Partial Hepatectomy Decreases Toxicity of CCl_4 in Gerbils," *Biochem. Pharmacol.* 42: 633–644 (1991).

123. Gerhard, H. J., B. Schultz, and W. Maurer. "Wirkung einer zweiten CCl_4-Intoxikation auf die CCl_4-geschadigte Leber de Mans," *Virchows Abk. B. Zellerpath.* 10:184–199 (1972).

124. Pound, A. W., and T. A. Lawson. "Reduction of Carbon Tetrachloride Toxicity by Prior Administration of a Single Small Dose in Mice and Rats," *Br. J. Exp. Pathol.* 56:172–179 (1975).

125. Klingensmith, S. J., and H. M. Mehendale. "Destruction of Hepatic Mixed Function Oxygenase Parameters by CCl_4 in Rats Following Acute Treatment with Chlordecone, Mirex and Phenobarbital," *Life Sci.* 33:2339–2348 (1984).

126. Klingensmith, S. J. "Metabolism of CCl_4 in Rats Pretreated with Chlordecone, Mirex and Phenobarbital," University of Mississippi Medical Center, PhD Dissertation (1982).

127. Thakore, K. N., and H. M. Mehendale. "Liver Injury and CCl_4 Autoprotection," *FASEB J.* 5:A1248 (1991).

128. Rao, C. V., and H. M. Mehendale. "Effect of Antimitotic Agent Colchicine on CCl_4 Toxicity," *Pharmacologist* 32:168 (1990).

129. Rao, C. V., and H. M. Mehendale. "Prolongation of Carbon Tetrachloride Toxicity by Colchicine Antimitosis," *Toxicologist* 11:128 (1991).

130. Rao, C. V., and H. M. Mehendale. "Effect of Colchicine on Hepatobiliary Function in CCl_4 Treated Rats," *Biochem. Pharmacol.* 43: in press (1991).

131. Rao, C. V., and H. M. Mehendale. "Colchicine Antimitosis Abolishes CCl_4 Autoprotection," *Toxicol. Pathol.* 19: in press (1991).

132. Faroon, O. M., and H. M. Mehendale. "Bromotrichloromethane Hepatotoxicity. Role of Hepatocellular Regeneration in Recovery. Biochemical and Histopathological Studies in Control and Chlordecone Pretreated Male Rats," *Toxicol. Pathol.* 18:667–677 (1990).

133. Faroon, O. M., R. W. Henry, M. G. Soni, and H. M. Mehendale. "Potentiation of $BrCCl_3$ Hepatotoxicity by Chlordecone: Biochemical and Ultrastructural Study," *Toxicol. Appl. Pharmacol.* 110: 185–197 (1991).

134. Thakore, K. N., M. L. Gargas, M. E. Andersen, and H. M. Mehendale. "PB-

PK Derived Metabolism Constants, Hepatotoxicity, and Lethality of BrCCl$_3$ in Rats Pretreated with Chlordecone, Phenobarbital and Mirex," *Toxicol. Appl. Pharmacol.* 109: 514–528 (1991).

135. Leevy, C. M., R. M. Hollister, R. Schmid, R. A. MacDonald, and C. S. Davidson. "Liver Regeneration in Experimental CCl$_4$ Intoxication," *Proc. Soc. Exp. Biol. Med.* 102:672–675 (1959).

136. Smuckler, E. A., M. Koplitz, and S. Sell. "A-Fetoprotein in Toxic Liver Injury," *Cancer Res.* 36:4558–4561 (1976).

137. Nakata, R., I. Tsukamoto, M. Miyoshi, and S. Kojo. "Liver Regeneration after CCl$_4$ Intoxication in the Rat," *Biochem. Pharmacol.* 34:586–588 (1985).

138. Ebel, R. E., and E. A. McGrath. "CCl$_4$-Hepatotoxicity in the Mongolian Gerbil: Influence of Monooxygenase Induction," *Toxicol. Lett.* 22:205–210 (1984).

139. Bhathal, P. S., N. R. Rose, I. R. Mackay, and S. Whittingham. "Strain Differences in Mice in Carbon Tetrachloride–Induced Liver Injury," *Br. J. Exp. Pathol.* 64:524–533 (1983).

140. Biesel, K. W., M. N. Ehrinpreis, P. S. Bhathal, I. R. Mackay, and N. R. Rose. "Genetics of Carbon Tetrachloride–Induced Liver Injury in Mice. II. Multigenic Regulation," *Br. J. Exp. Pathol.* 65:125–131 (1984).

141. McLean, M. R., and K. R. Rees. "Hyperplacia of Bile Ducts Induced by Alpha-naphthyl-iso-thiocyanate: Experimental Biliary Cirrhosis Free from Obstruction," *J. Pathol. Bacteriol.* 76:175–188 (1958).

142. Ungar, H., E. Moran, M. Eisner, and M. Eliakim. "Rat Intrahepatic Biliary Tract Lesions from Alpha-naphthyl Isothiocyanate," *Arch. Pathol.* 73:427–435 (1962).

143. Condie, L. W., C. L. Smallwood, and R. D. Laurie. "Comparative Renal and Hepatotoxicity of Halomethanes: Bromodichloromethane, Bromoform, Chloroform, Dibromochloromethane and Methylene Chloride," *Drug Chem. Toxicol.* 6:563–578 (1983).

144. Natchtomi, E., and E. Farber. "Ethylene Dibromide as a Mitogen for Liver," *Lab. Invest.* 38:279–283 (1978).

145. Lesch, R., W. Reutter, D. Keppler, and K. Decker. "Liver Restitution after Galactosamine Hepatitis: Autoradiographic and Biochemical Studies in Rats," *Exp. Mol. Pathol.* 12:58–69 (1970).

146. Kuhlmann, W. D., and K. Wurster. "Correlation of Histology and Alpha-fetoprotein Surgence in Rat Liver Regeneration after Experimental Injury by Galactosamine," *Virchows Arch. Histol.* 387:47–57 (1980).

147. Gupta, D. N. "Acute Changes in the Liver after Administration of Thioacetamide," *J. Pathol. Bacteriol.* 72:183–192 (1956).

148. Reddy, J. K., M. Chiga, and D. Svoboda. "Initiation of Division of Cycle of Rat Hepatocytes Following a Single Injection of Thioacetamide," *Lab. Invest.* 20:405–411 (1969).

CHAPTER 5

Effects of Low-Dose Radiation
on the Immune Response

Robert E. Anderson, University of New Mexico, School of Medicine, Albuquerque, New Mexico

The extreme radiosensitivity of the small lymphocyte has been known for some time.[1-10] More recently, it has been recognized that under highly specific experimental conditions, ionizing radiation may also exert an immunostimulatory effect. The purpose of this chapter is to summarize our current understanding of this phenomenon with particular reference to

1. radiation dose
2. the involved cell type(s)
3. the temporal relationships between exposure to radiation, introduction of antigen, or related immunostimulatory agent, and the resultant immune response

Prior to reviewing the experimental data, however, a few introductory comments with respect to the normal function of the immune system are in order (for a more detailed description, see Anderson et al.[11]).

IMMUNE SYSTEM

The raison d'être of the immune system is to protect the host against infectious agents and their products. In order to discharge this function, the various involved cells and cell types are in a constant state of flux as the host responds to a continuous series of both new and familiar stimuli, while simultaneously experiencing the slow decay of responses to more remote stimuli. Thus, at any one time, a myriad of reactions are commencing while others are reaching their maximum and still others are waning.

The immediate response of the host to an unfamiliar immunologic stimulus involves a complex series of events encompassing several different types of cells located in a variety of anatomic locations. These events are regulated by a complex communications network, involving not only several distinct cell types but also a potpourri of factors that they produce. The involved cells are often grouped together in highly specific spatial relationships in a series of so-called lymphoreticular tissues, which include the thymus, spleen, lymph nodes, and the gastrointestinal tract–associated lymphoid tissues. In addition, one of the cell types—the lymphocyte—constantly recirculates throughout the host in highly specific fashion, as determined by a complex of cell surface molecules and their tissue equivalents.

The major cellular components of the immune system are lymphocytes, macrophages, and a series of macrophage-related accessory cells. Lymphocytes are of critical importance in almost every facet of the immune response. Among their functions are the recognition of antigen as foreign ("non-self"), the induction and modulation of the resultant immune response, and, most particularly, the high degree of specificity of these responses. Lymphocytes are also responsible for the "memory" that accompanies most immune responses and the "tolerance" that prevents host cells from initiating a response directed against host antigens.

The role of the macrophage is less specific but no less critical than that of the lymphocyte. Whereas individual lymphocytes are precommitted to respond to a limited number of structurally related antigens, macrophages appear to be far less discriminating. Thus, at present, they are not thought to have the same high degree of antigen specificity that characterizes lymphocytes. Rather, macrophages are responsible for the phagocytosis and degradation of complex antigens to the more simplified forms that are then susceptible to recognition by lymphocytes. Macrophages also produce a number of immunologically active mediators that are capable of regulating various types of lymphocytes.

Accessory cells resemble macrophages in many ways but appear to be functionally more sophisticated. Perhaps as a consequence, accessory cells are often found in highly specific locations in complex lymphoreticular tissues such as the thymus.

Despite morphologic homogeneity, at least at the light microscope level, small lymphocytes may be subdivided functionally into two major groups:

1. *Thymus-derived or T cells.* These cells are primarily responsible for cellular immunity and delayed-type hypersensitivity responses (transplantation immunity and immunity to select viral, bacterial, and parasitic antigens).
2. *Bone marrow–derived or B cells.* These lymphocytes are primarily responsible for humoral immunity and the production of circulating antibody; under the influence of antigen, B cells can differentiate into plasma cells.

Both T and B cells can be further subdivided on the basis of cell surface markers and specific immune functions (Table 5.1).[11] Of particular importance, in the context of this chapter, is the observation that there are a number of subsets of T cells that serve to modulate the effects of other T, as

Table 5.1. Select Characteristics of T and B Cells

	T Cells	B Cells
Cell of origin	Hematopoietic stem cell	Hematopoietic stem cell
Site of maturation	Thymus	Bursa Fabricius (birds); bone marrow (mammals)
Function	Regulatory (help, suppression) and effector (delayed-type hypersensitivity, mixed lymphocyte reaction, cytotoxic or "killer")	Precursor of antibody-producing cell
Primary anatomic localization		
—Spleen	Periarteriolar cuff	Follicles
—Lymph node	Para-cortex	Primary follicles
—Peyer's patch	Interfollicular	Follicles
Recirculating pool	Majority (80–85% in mouse)	Minority (15–20% in mouse)
Life span	4–6 months	6–8 weeks
Mitogenic response	PHA, Con A	LPS Purified protein derivative
Memory	Yes	Yes

Source: After Anderson et al.[11]

Table 5.2. Necrosis vs Apoptosis

	Necrosis	Apoptosis
Injurious insult	+ (hypoxia, toxins, etc.)	— (programmed)
Geography	Local/focal	Semirandom
Inflammatory response	+	—
Morphology —Nuclear chromatin	Flocculates (late)	Condenses, fragments (early)
—Cytoplasm	Swells, blebs develop	Condenses, organelles preserved
Executioner	Ca^{++}-dependent phospholipases	Ca^{++}-dependent endonuclease

well as B, cells. The two principal categories of regulatory T cells are appropriately termed "helper" and "suppressor" cells.

T cells, and possibly B cells, are also unusual in that the vast majority are destroyed prior to their complete differentiation and release to the peripheral lymphoid tissues of the host.[11] With T cells, this phenomenon (known as *apoptosis*) occurs within the thymus and results in the death in situ of approximately 90% of developing thymocytes. A similar phenomenon has been suspected, but as yet not documented, for B cells. In part, failure to demonstrate this phenomenon with B cells may relate to the observation that in most rodents, and probably also in humans, the differentiation of B cells occurs in the same anatomic site as does the generation of pre-B cells (i.e., the bone marrow), which markedly confounds the interpretation of experiments designed to address this issue.

RADIATION

The cell types involved in the immune response exhibit a broad spectrum of radiosensitivities.[9] For example, some subpopulations of lymphocytes are exceedingly radiosensitive, while plasma cells and macrophages are very resistant. The basis of these differences in radiosensitivity are not well understood, but presumably relate to the degree of differentiation and maturation of the involved cell type, as well as to whether or not cell division is a requisite for participation. In addition, some types of lymphocytes die acutely after irradiation (so-called interphase cell death) while others succumb in more traditional fashion. Interphase cell death is of particular importance with low-dose exposures and may, in fact, account for the exquisite sensitivity of the small lymphocyte to low doses of radiation. Interphase cell death is distinctive morphologically and is similar to, if not identical with, apoptosis.[13-15] The features that distinguish apoptosis/interphase death from necrosis are listed in Table 5.2.

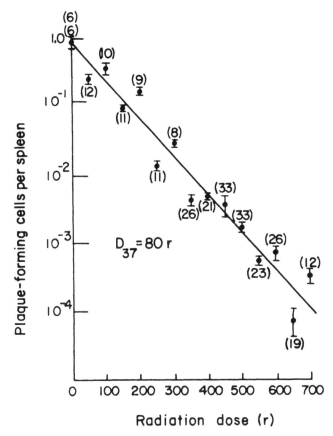

Figure 5.1. Relationship between magnitude of irradiation and plaque-forming capacity of mice. SRBC, 4×10^8, were given 10 days after whole-body irradiation, and assays for plaque-forming spleen cells were performed 4 days later. The results are plotted relative to the plaque-forming response of control (nonirradiated) mice. Adapted from Kennedy et al.[3]

In general, irradiation inhibits the immune response in a dose-dependent fashion. This is illustrated in Figure 5.1, taken from a study by Kennedy et al.[3] Mice were exposed to the indicated doses of radiation, injected with antigen 10 days later, and then assayed for their capacity to produce antibody. Similar results have been obtained in vitro,[16] demonstrating that the radiation-induced loss of functional activity is due to injury of the involved cell types rather than a perturbation of the environment provided by the host.

Whole body irradiation of mice also results in a dose-dependent loss of substance among the various lymphoreticular tissues.[9] As shown in Table 5.3 and Figure 5.2, this relationship is reflected in the recirculation pool.[17] Table 5.3 shows the number of lymphocytes mobilizable by thoracic duct cannulation as a function of radiation dose at various time intervals up to

Table 5.3. Number of Thoracic Duct Lymphocytes as a Function of Radiation Dose and Time After Irradiation

Radiation Dose (rads)	No. of Mice	4 Hours				10 Hours				21 Hours				46 Hours			
		No. of Cells	Viability	Percent T	Percent B	No. of Cells	Viability	Percent T	Percent B	No. of Cells	Viability	Percent T	Percent B	No. of Cells	Viability	Percent T	Percent B
0	4	11.8	100	65	33	25.6	98	58	39	33.5	99	56	43	33.8	99	54	48
50	4	9.8	99	63	35	14.5	99	56	44	18.7	99	52	48	7.4	99	53	48
100	8	6.8	99	89	11	8.8	99	90	9	6.1	95	85	11	6.1	95	86	14
300	4	9.5	99	87	9	3.7	96	93	7	6.4	99	93	9	4.6	99	92	6
500	5	3.8	99	59	35	4.8	100	58	40	3.3	100	52	47	1.2	99	55	41

Source: After Anderson and Williams. [17]

Note: Four-month-old CBA female mice were exposed in whole-body fashion to indicated dose. Immediately after irradiation the thoracic duct was cannulated, and the number of TDL determined at the indicated times. Only mice which flowed well were employed for data calculations. Results represent cumulative cell numbers.

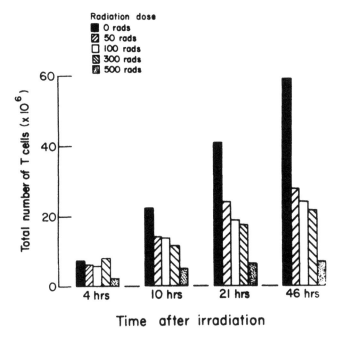

Figure 5.2.   Cumulative number of viable T cells mobilizable by thoracic duct cannulation as a function of radiation dose and time after exposure. Data derived from Table 5.3. Adapted from Anderson and Williams.[17]

46 hr after exposure. After the initial 4 hr, a dose-dependent decrement in the number of thoracic duct lymphocytes (TDLs) is apparent. As shown in Figure 5.2, this decrease is primarily related to a loss of recirculating T cells.

Recovery of lymphoreticular tissues after irradiation is much more complex than might be anticipated. In part, this complexity relates to the observation that the restoration of some cell types requires the expansion of a precursor pool anatomically removed from the tissue of interest. In addition, the lymphoreticular tissues themselves have a hierarchy of recovery, and regeneration of so-called "peripheral" organs, such as the spleen and lymph nodes, is in part dependent upon recovery of the "central" thymus. Some of these interrelationships are illustrated in Figures 5.3 and 5.4.[18]

Although the immunosuppressive effects of ionizing radiation have been recognized for some time, more recently it has been shown that such exposures may also occasion an augmented response.[19] This immunostimulatory effect is dependent particularly on the dose of radiation and the time between exposure and the introduction of antigen. Much of the early experimental work concerned with this phenomenon was performed by Dixon and Taliaferro and their respective co-workers.[5,6,20–22] Figures 5.5 and 5.6, from Taliaferro et al.,[5,6] show the effects of small versus large amounts of whole

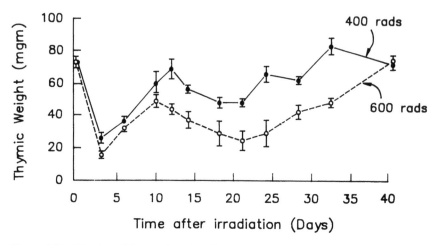

Figure 5.3. Thymic weight as a function of dose and time after irradiation. Adapted from Anderson and Hendry.[18]

body irradiation administered to rabbits at various times before or after primary (Figure 5.5) or secondary (Figure 5.6) immunization with sheep erythrocytes. As noted previously, the temporal relationship between irradiation and antigen injection appears critical to the generation of an augmented response. Also important, although apparently not as critical as timing, are the character of the antigen; the radiation dose; the manner in which the irradiation is administered; and the age, sex, and genetic composition of the host.

Low-dose radiation can also be utilized to immunize animals against

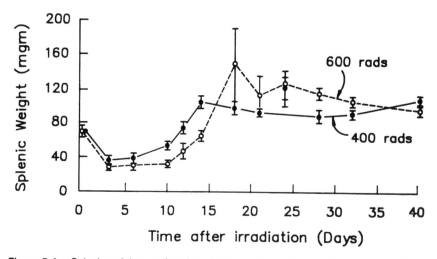

Figure 5.4. Splenic weight as a function of dose and time after irradiation. Adapted from Anderson and Hendry.[18]

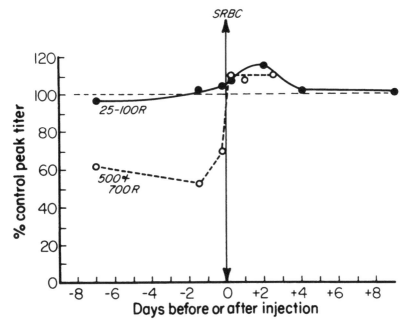

Figure 5.5.   Comparison of effects of small versus large amounts of whole-body irradiation administered to rabbits at various times before or after primary immunization with SRBC. Adapted from Taliaferro and Taliaferro.[5]

amounts of antigen that would otherwise be subimmunogenic.[12] In nonirradiated mice, such "low-dose tolerance" is apparently due to the induction of suppressor T cells. Figure 5.7 shows the effect of whole body exposure to small amounts (15 rad) on the response of mice to inactivated (mitomycin-treated) tumor cells. The mice were irradiated or sham-irradiated immediately prior to injection with the indicated numbers of mitomycin-treated tumor cells. A control group was given phosphate-buffered saline. After 21 days, all mice received 10^4 viable tumor cells. As seen in the figure, sham-irradiated mice injected with small numbers of mitomycin-treated Sarcoma I (SaI) cells exhibited larger tumors than the saline-injected controls (solid line) when subsequently challenged with untreated tumor cells. Exposure to low-dose irradiation not only abolished this partial tolerance to the tumor cells, but actually rendered the irradiated mice partially immune. The augmented antitumor activity of these (low-dose) irradiated spleen cells can also be shown in a cell transfer system (Figure 5.8).[23]

The above experiments have involved rabbits and mice irradiated in vivo. A similar phenomenon can be demonstrated in vitro. The latter approach has the added advantage of permitting the dissection of the various components of the immune response by subtracting or adding specific cell types. Figure 5.9 shows the response of murine spleen cells exposed to several doses of radiation in vitro to sheep red blood cells (SRBCs) as antigen.[16] The

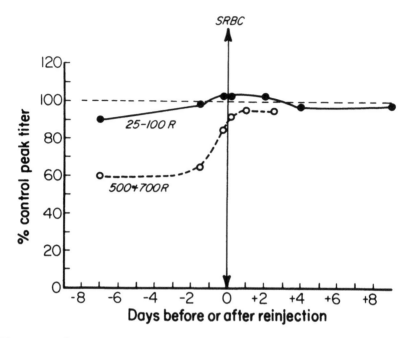

Figure 5.6. Comparison of effects of small versus large amounts of whole-body irradiation administered to rabbits at various times before or after secondary immunization with SRBC. Adapted from Taliaferro and Taliaferro.[6]

results are expressed both in absolute terms (dashed line) and as a percent of the control response (shaded areas). An augmentation of the response is clearly evident with low-dose exposures (5–50 rad). This radiation-induced augmentation is associated with reduced numbers of viable cells (Figure 5.10), suggesting the inhibition or death of a cell responsible for modulating the response.

In order to determine the cell type involved in low-dose augmentation, the immune response was divided into its T and B cell components, and each was irradiated individually and then recombined with the corresponding (nonirradiated) cell type. Results from this approach are shown in Figures 5.11 and 5.12.[16] Figure 5.11 shows the effect of irradiation of the B cell component on the in vitro response to SRBCs. The resultant dose-response curve shows three apparent components:

1. equivocal augmentation associated with low (0–20 rad) doses
2. marked loss of activity associated with moderate (25–75 rad) doses
3. minimal additional loss of activity associated with large (≥ 100 rad) doses

A complementary experiment utilizing irradiated T cells is shown in Figure 5.12. This dose-response curve similarly shows three components but differs from Figure 5.11 as follows:

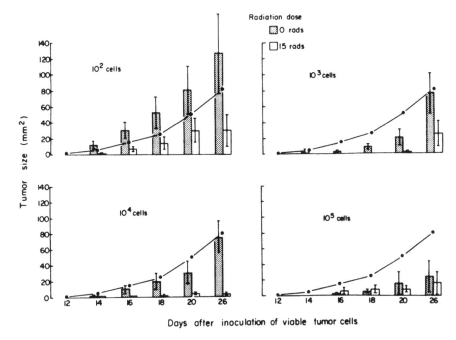

Figure 5.7. Effect of 15 rads upon response of A/J mice to varying numbers of mitomycin-treated Sal cells. Groups of 20 mice were exposed to 15 rads whole-body irradiation or sham-irradiated, and inoculated subcutaneously with the indicated numbers of mitomycin-treated tumor cells. Twenty-one days later, all animals received 10^4 untreated Sal cells and were followed for tumor size. A control group *(solid line)* did not receive mitomycin-treated cells. Adapted from Anderson et al.[12]

1. The augmentation is much more pronounced and appears to be unequivocal.
2. The doses associated with each phase are distinctly different.

The introduction of antigen, either in vivo or in vitro, results in the rapid expansion of the corresponding subsets of lymphocytes. This entails the rapid division of the involved cell types. Cell division may also be occasioned by the introduction of a group of related plant derivatives termed *mitogens.*

In the context of radiation injury of the immune response, mitogens have the added advantage of selectively stimulating specific subpopulations of lymphocytes. Phytohemagglutinin (PHA), for example, stimulates both helper and suppressor T cells, Concanavalin A (Con A) activates proportionately more suppressor cells, and lipopolysaccharide (LPS) activates B cells. Figure 5.13 shows the effect of various doses of radiation on the in vitro response of mouse spleen cells to the indicated mitogens.[24] Augmented responses are seen with both PHA and Con A.

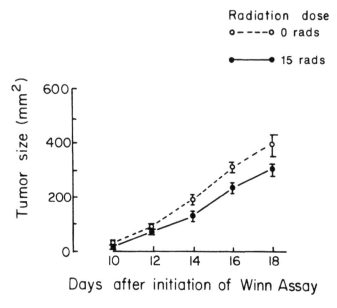

Figure 5.8.   Effect of in vitro irradiation (15 rads) of donor spleen cells upon tumor size in a Winn assay. Mice were inoculated subcutaneously with 10^4 Sal cells; after 2 days the mice were killed, and their spleens irradiated or sham-irradiated and then used in a Winn assay with an equal number of Sal cells. Results represent mean tumor area ± S.E. Adapted from Anderson et al.[23]

DISCUSSION

Under appropriate experimental conditions, radiation exposure can be associated with augmentation of the immune response. This phenomenon can be demonstrated both in vivo and in vitro and appears to be due to the loss of a T cell that under normal circumstances exerts a suppressive influence. According to this hypothesis, radiation-induced injury to this cell type causes a loss of suppression and thereby results in augmentation.

As a part of normal differentiation, T cells undergo spontaneous cell death within the thymus. This process, known as apoptosis, is thought to be responsible for the elimination of cells with the potential of reacting against the host and thereby eliciting an autoimmune reaction. The morphological similarities between apoptosis and radiation-induced interphase cell death are striking and suggest the possibility of one or several common denominators.

Apoptosis is an energy-dependent phenomenon characterized by the condensation of nuclear chromatin and the fragmentation of DNA at internucleosomal linker sites. As a consequence, electrophoresis of DNA obtained from cells undergoing apoptosis yields discrete bands with multiples of 180 to 200 base pairs.[25] In normal thymus cells, this phenomenon has been shown to be due to the activation of a calcium-dependent endonuclease.[26]

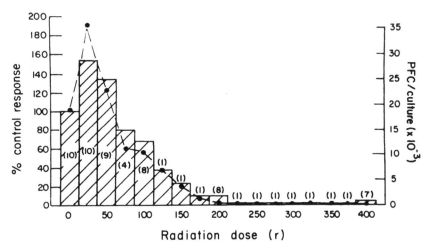

Figure 5.9. Effect of irradiation upon anti-SRBC response. C57BL/6 spleen cells from female donors were exposed to indicated doses of radiation in vitro and incubated in 1-mL cultures with SRBCs as antigen. The number of anti-SRBC plaque-forming cells (PFCs) was determined on day 4. The results are expressed as a percentage of the control (0 rad) response. The numbers in parentheses indicate the number of experiments included in the calculation of the indicated value. The actual data from a single experiment *(dotted line)* are included for comparative purposes. Adapted from Anderson and Lefkovits.[16]

Although the intrathymic deletion of T cells probably is genetically initiated, perhaps by permitting an influx of calcium, radiation-induced interphase death is clearly triggered by the physical event per se. The end result, interphase death, may be due to either a direct or an indirect effect, such as free radical formation. Whatever the initial event, however, it is easy to envision a series of steps that culminate in endonuclease activation.

As noted previously, low-dose augmentation of the immune response appears to relate to the functional loss of an immunoregulatory cell of the T cell lineage. The requisite temporal relationships among irradiation, introduction of antigen (or mitogen), and an augmented response suggest that this phenomenon may be due to interphase cell death similar, at least morphologically, to that observed among normal T cells during their differentiation within the thymus. Although spontaneous and radiation-initiated apoptosis have been particularly well studied in the thymus, it must be reemphasized that the phenomenon is not limited to T cells. In fact, a recent report documents the same phenomenon in B cells.[27]

Is there an evolutionary basis for the extreme radiosensitivity of some types of lymphocytes? Does it offer some type of competitive advantage to the host? Several hypothetical arguments can be advanced:

1. *Coincidence.* It is also possible that the relationship between apoptosis and radiation-induced interphase death is purely coincidental, that radiation

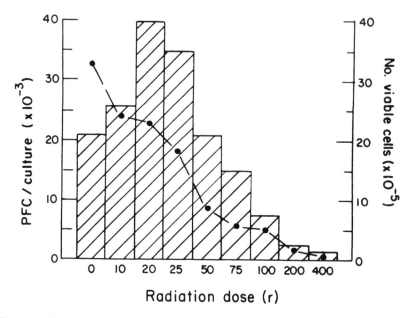

Figure 5.10. Relationship between cell viability and radiation-induced augmentation. Spleen cell suspensions were irradiated in vitro and placed in cultures with SRBCs as antigen. Results represent average of two experiments harvested on day 4. Similar results (not shown) were obtained with analyses on day 5. Adapted from Anderson and Lefkovits.[16]

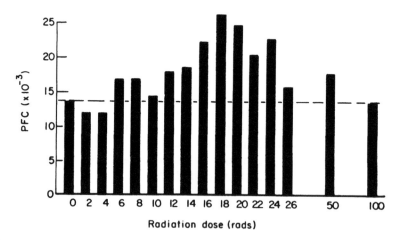

Figure 5.11. Influence of small increments of irradiation upon in vitro response to SRBCs. Spleen cells were irradiated in vitro and placed in 1-mL cultures with SRBCs as antigen. Data represent plaque-forming cells (PFCs) on day 4. Similar results obtained on day 5 (data not shown). Adapted from Anderson and Lefkovits.[16]

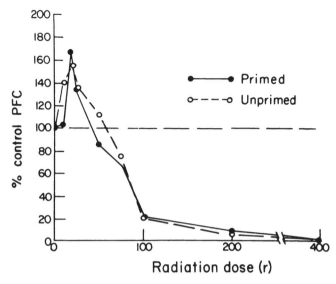

Figure 5.12. Influence of irradiation upon primed versus unprimed cells. Mice were injected with SRBCs or saline 40 days prior to sacrifice. Spleen cell suspensions were exposed to indicated doses of radiation in vitro and placed in 1-mL cultures. Results expressed as a percentage of the control response. Adapted from Anderson and Lefkovits.[16]

injury initiates the "apoptotic pathway" at a point subsequent to the initiating intrathymic event. Thus, it is entirely a chance phenomenon that the end result is the same.

2. *Competitive advantage.* Since all forms of life are continuously exposed to various forms of "natural" or background radiation, perhaps low-dose exposure offers a competitive advantage via nonspecific augmentation of the immune response. The resting state, according to this line of thinking, is really not resting but rather one of heightened reactivity as a consequence of background exposures. This proposition seems unlikely, especially since the doses associated with augmentation experimentally are several logs larger than those experienced from natural exposures in most parts of the world.

3. *Curb against autoimmunity.* It is clear that the mammalian host goes to great lengths to protect itself against autoreactive lymphocytes. In particular, the passage of pre-T cells through the thymus is accompanied by the loss of most of these cells, presumably to protect the host against the release and peripheralization of cells with self-reactive potential. In this context, it is important to note that the lymphocyte is the only mammalian cell that can undergo extensive clonal expansion postnatally. As a consequence, one renegade autoreactive lymphocyte could in theory proliferate in response to histoincompatible host antigens, thereby developing a self-reactive clone capable of killing the host. Similarly, a lymphocyte that is injured but not killed by irradiation poses a distinct threat to the host, especially if the damage involves the nucleus. Nonlethal damage to lymphocyte DNA could result in the mutation of genes associated with regula-

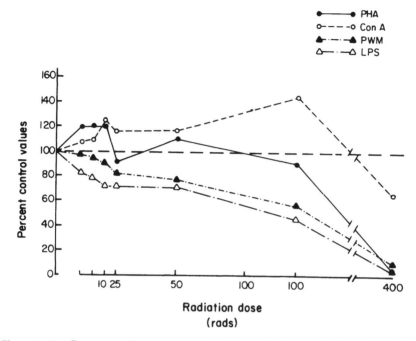

Figure 5.13.   Response of irradiated spleen cells to optimal concentrations of PHA, Con A, pokeweed mitogen (PWM), and LPS. Results are given as a percentage of the corresponding control (0 rad) values *(broken line)*. Adapted from Anderson and Troup.[24]

tion of growth (and tumorigenesis) or those involved in definition of "self" (and thus autoreactivity). This line of thinking suggests that it is in the host's best interests to do away with irradiated lymphocytes as quickly (interphase) and as efficiently (apoptosis) as possible.

Of these possibilities, the third is clearly the most attractive, in part because it is susceptible to testing experimentally.

REFERENCES

1. Trowell, O. A. "The Sensitivity of Lymphocytes to Ionizing Radiation," *J. Pathol.* 64:687 (1952).
2. Makinodan, T., M. A. Kastenbaum, and W. J. Peterson. "Radiosensitivity of Spleen Cells from Normal and Preimmunized Mice and Its Significance to Intact Animals," *J. Immunol.* 88:31 (1962).
3. Kennedy, J. C., J. E. Till, L. Simonovitch, and E. A. McCullough. "Radiosensitivity of the Immune Response to Sheep Red Cells in the Mouse, As Measured by the Hemolytic Plaque Method," *J. Immunol.* 94:715 (1965).
4. Janeway, C. A. "Cellular Cooperation During In Vivo Antihapten Antibody Responses. II. The Effect of In Vivo and In Vitro X-Irradiation on T and B Cells," *J. Immunol.* 114:1402 (1975).

5. Taliaferro, W. H., and L. G. Taliaferro. "Effects of Radiation on the Initial and Anamnestic IgM Hemolysin Responses in Rabbits: Antigen Injection after X-rays," *J. Immunol.* 103:559 (1969).

6. Taliaferro, W. H., and L. G. Taliaferro. "Effects of Irradiation on Initial and Anamnestic Hemolysin Responses in Rabbits: Antigen Injection before X-rays," *J. Immunol.* 104:1364 (1970).

7. Hamaoka, T., D. H. Katz, and B. Benacerraf. "Radioresistance of Carrier-Specific Helper Thymus-Derived Lymphocytes in Mice," *Proc. Nat. Acad. Sci.* 69:3453 (1972).

8. Micklem, H. S., and J. F. Loutit. *Tissue Grafting and Radiation* (New York: Academic Press, 1966).

9. Anderson, R. E., and N. L . Warner. "Ionizing Radiation and the Immune Response," *Adv. Immunol.* 24:215 (1976).

10. Hellström, K. E., I. Hellström, J. A. Kant, and J. D. Tamerius. "Regression and Inhibition of Sarcoma Growth by Interference with a Radiosensitive T-Cell Population," *J. Exp. Med.* 148:799 (1978).

11. Anderson, R. E., J. C. Standefer, and S. Tokuda. "The Structural and Functional Assessment of Cytotoxic Injury of the Immune System with Particular Reference to the Effects of Ionizing Radiation and Cyclophosphamide," *Brit. J. Cancer* 53(Suppl. VII):140 (1986).

12. Anderson, R. E., S. Tokuda, W. L. Williams, and C. W. Spellman. "Low Dose Irradiation Permits Immunization of A/J Mice with Subimmunogenic Numbers of SaI Cells," *Brit. J. Cancer* 54:505 (1986).

13. Wyllie, A. H., J. F. R. Kerr, and A. R. Currie. "Cell Death: The Significance of Apoptosis," *Int. Rev. Cytol.* 68:251 (1980).

14. Wyllie, A. H., R. G. Morris, A. L. Smith, and D. Dunlop. "Chromatin Cleavage in Apoptosis: Association with Condensed Chromatin Morphology and Dependence on Macromolecular Synthesis," *J. Pathol.* 142:67 (1984).

15. Duvall, E., and A. H. Wyllie. "Death and the Cell," *Immunol. Today* 7:115 (1986).

16. Anderson, R. E., and I. Lefkovits. "In Vitro Evaluation of Radiation-Induced Augmentation of the Immune Response," *Am. J. Path.* 97:456 (1979).

17. Anderson, R. E., and W. L. Williams. "Radiosensitivity of T and B Lymphocytes: V. Effects of Whole Body Irradiation on Numbers of Recirculating T Cells Sensitized to Primary Skin Grafts in Mice," *Am. J. Path.* 89:367 (1977).

18. Anderson, R. E., and J. Hendry. Unpublished data (1986).

19. Anderson, R. E., I. Lefkovits, and G. M. Troup. "Radiation-Induced Augmentation of the Immune Response," *Contemporary Topics in Immunobiology,* Vol. 11, N. L. Warner, Ed. (New York: Plenum Publishing, 1980), pp. 245–274.

20. Taliaferro, W. H., L. G. Taliaferro, and B. N. Jaroslow. "Radiation and Immune Mechanisms," (New York: Academic Press, 1964).

21. Dixon, F. J., and P. J. McConahey. "Enhancement of Antibody Formation by Whole Body X-Irradiation," *J. Exp. Med.* 117:833 (1963).

22. Schmidtke, J. R., and F. J. Dixon. "Effects of Sublethal Irradiation on the Plaque-Forming Cell Response in Mice," *J. Immunol.* 111:691 (1973).

23. Anderson, R. E., W. L. Williams, and S. Tokuda. "Effect of Low Dose Irradiation upon T Cell Subsets Involved in the Response of Primed A/J Mice to SaI Cells," *Int. J. Radiat. Biol.* 53:103 (1988).

24. Anderson, R. E., and G. M. Troup. "Effects of Irradiation upon the Response of Murine Spleen Cells to Mitogens," *Am. J. Pathol.* 109:169 (1982).

25. Wyllie, A. H. "Glucocorticoid-Induced Thymocyte, Apoptosis Is Associated with Endogenous Endonuclease Activation," *Nature* 284:555 (1980).

26. Cohen, J. J., and R. C. Duke. "Glucocorticoid Activation of a Calcium-Dependent Endonuclease in Thymocyte Nuclei Leads to Cell Death," *J. Immunol.* 132:38 (1984).

27. Fry, A. M., L. A. Jones, A. M. Kruisbeck, and L. A. Matis. "Thymic Requirement for Clonal Deletion During T Cell Development," *Science* 246:1044 (1989).

CHAPTER 6

Cellular Adaptation as an Important Response During Chemical Carcinogenesis

Emmanuel Farber, Departments of Pathology and of Biochemistry, University of Toronto, Ontario, Canada

INTRODUCTION

All living forms have evolved in a largely unfriendly, or even hostile, environment in which protective responses to many different forms of potential injury or harm have been essential for survival and reproduction. The acquisition of mechanisms for many different adaptive responses could be considered to be just as important as the development of the essential pathways for the basic physiological needs, such as for energy and for the syntheses of small and large molecules that are the essential components of the pathways. It is widely recognized that a variety of physiological adaptive responses to varying altitudes, other environmental influences, and hormonal modulations, as well as aging, are fundamental properties of living organisms. In addition, the vast array of different xenobiotic chemicals and organisms as well as radiations to which all living forms have been exposed since the early forms evolved some 2 to 3 billion years ago would require highly versatile protective systems to be developed. Such protective or adaptive mechanisms are known to be present throughout the whole spectrum of living forms — from single cell microorganisms to highly differentiated eukaryotes.

Since disease processes are largely expressions of how living organisms react and respond to perturbations in the external and internal environments, adaptive or protective responses and their modulations and mechanisms are of the greatest concern in fundamental studies of disease pathogenesis. Such considerations are also of the greatest relevance in toxicology, including how living organisms respond to low levels of single and multiple xenobiotics and radiations.

In the face of the many hazards, including carcinogenic ones, it seems appropriate to ask the question — what mechanisms might have evolved to preserve biological continuity? In species that have survived and endured over the eons, what is the nature of the adaptations that have evolved, and could a knowledge of these be useful in designing new approaches to the prevention of cancer, and perhaps even its treatment? What essential roles, if any, do these adaptive processes play in the development of different cancers?

CONCEPTS OF CANCER AND CANCER DEVELOPMENT

Fundamentally, there have been two major concepts that have guided the majority of cancer researchers. Scientists and clinicians have discussed for decades the issue of whether cancer is fundamentally a genetic or better "genomic disease," or is it more likely to be an "epigenetic process." A large component of the latter could be considered under the rubric of "adaptive responses."

During the past 20 years, the evidence in favor of an interaction of genotoxic carcinogenic chemicals, radiations, DNA viruses, and RNA viruses with the genome of the target cells is so overwhelming that one cannot seriously consider cancer development as a biological process not involving the genome as a critical component. Even considerations of differentiation and cancer, a most interesting area, cannot deny genomic changes as crucial to the carcinogenic process. Mutations, gene rearrangements, translocations, and/or other forms of genomic disorganization are common accompaniments of cancer.

Yet, we are faced with an increasing realization that there is an expanding number of "nongenotoxic" chemicals, such as several hypolipidemic and other drugs, and some halogenated hydrocarbons used in industry and agriculture, that have to be studied. These are unquestionably carcinogenic, yet do not seem to show either the interactions with DNA, including mutagenicity, in many prokaryotes and eukaryotes or the well-delineated initiation steps seen readily with the genotoxic carcinogens.

In addition, as the steps and mechanisms during cancer development are studied in greater depth, phenomena become apparent that suggest that adaptive reactions and responses may play important or even critical roles in the process of carcinogenesis.

Given the acceptance of some types of genomic changes in many known instances of cancer development, we are faced with a much more important as well as a much more practical consideration. Do the cells with altered genomes behave essentially in a confrontational role in the host to create an adversarial situation, or do the selective and other important host processes exercise major options and limits to the type of initiated cell allowed to persist and grow? In other words, is the process of carcinogenesis fundamentally an adversarial one (i.e., an abnormal cell in a vulnerable host), or is it more in the nature of a physiological selection or differentiation, a form of metaplasia, which has survival value for the host as an adaptive phenomenon?

Adversarial-Confrontational

The overwhelming view of both scientists and clinicians is biased toward the first possibility, the adversarial. This view naturally considers the approach to cancer therapy as one of creating more efficient ways to *kill* the abnormal cells. This approach is not only used for cancer but also for cancer precursors. Included, of course, are the older and newer approaches to the immunological control of cancer.

Since the cancer cells and their precursors are assumed to be "abnormal" in that they are, at least in part, foreign to the host, it is natural that immunological approaches be encouraged, including the use of killer lymphocytes and other categories of lymphocytes and cytotoxic macrophages.

This dominant monolithic view of cancer cells and their precursors in the carcinogenic process is a natural outcome of the nature of the current emphasis in cancer research. Since a major goal today of most cancer research is the cure of endstage cancer, and since some cancers can be initiated by known carcinogens, it is to be expected that the focus in cancer research should be on the "end," the out-and-out cancer and its behavior, and to a lesser degree on the beginning. The very long process whereby cancer develops, the carcinogenic process, is of necessity largely ignored, since it is only visible and easily amenable to treatment in a few types of cancers in humans, such as malignant melanoma and cervical carcinoma.

It is only natural that the cancer cell should be viewed as an abnormal or foreign cell that must be eradicated, given the following:

1. the genotoxic nature of many chemical carcinogens, radiations, DNA viruses, and probably RNA viruses
2. the wide range of genomic alterations readily induced by many of these agents
3. the obvious "physiologically abnormal" behavior of cancer cells and their progeny
4. the many genomic alterations seen in most cancers
5. the hereditary behavior of the cancer phenotype
6. the genomic disorganization so common in most, if not all, cancers
7. the extreme diversity and heterogeneity of cancers, even of a single cell type

Thus, the adversarial or confrontational view of cancer and cancer development has a considerable justification.

Adaptive-Physiological

Since the evidence in support of the adversarial-confrontational concept of cancer development is considerable, why should an alternative view, an adaptive one, be given serious consideration? The scientific justification for such a different view of the nature of the responses of the living eukaryote to carcinogens has been discussed in the past few years[1-5] and will only be briefly presented in this chapter.

What evidence is there for a physiological-adaptive view for carcinogenesis?

1. The earliest new cell populations that appear after initiation show a common phenotype in the liver regardless of the chemical nature and pattern of metabolism of the chemical carcinogen.[3,6]
2. This constitutive new phenotype in the rare altered hepatocyte is very similar to that induced reversibly in the whole liver by BHA, BHT, lead nitrate, or an interferon.[4,7] Thus, the new phenotype in the rare altered cell is not abnormal but can be "turned on" by other environmental perturbations. It consists of many enzymes and proteins.
3. A seemingly similar phenotype is induced by exposure of liver epithelial cell cultures to two retroviral oncogenes and a chemical carcinogen in vitro.[8,9]
4. The new cell populations that are induced by carcinogens are clearly a new pattern of differentiation with at least two biological options: differentiation to the mature adult liver as the major option and persistence with slow progression to cancer as a minor one.[3,10] The differentiation option is clearly genetically programmed since it occurs spontaneously and involves many enzymes and proteins, cell structure, cell to cell organization, and the blood supply, as well as other physiological parameters.
5. The clonal expansion during promotion of carcinogen-induced rare hepatocytes with a resistance phenotype, producing a liver with many nodules of resistant hepatocytes, is associated with an obvious protective role in the liver and for the host against cytotoxic and lethal effects of some xenobiotics, including carcinogens.[3,11-17]
6. Hepatocyte proliferation in the putative precancerous expanded clones, the persistent hepatocyte nodules, is almost balanced by hepatocyte cell loss in these nodules until late in the carcinogenic process when unequivocal cancer appears.[18-20] Until malignant neoplastic changes appear, the nodules grow very slowly. The balance between cell proliferation and cell loss is a physiological feature of the normal liver when the liver is exposed to primary mitogens that induce hyperplasia over and above the normal physiological size of that liver.[21,22] Thus, the nodules retain a major physiological control for cell proliferation until very late in the process of cancer development.
7. There appears to be no immune response to the new cell population until very late in the carcinogenic process with the final progression to cancer.[23] The late immunologic responses might be related to the common occurrence in virtually all cancers, but especially those with epithelial origin (carcinomas), of cell death with inflammation and the release of probably hundreds of cell constituents that normally never leave the cell until it dies.

In view of the many different cellular and tissue biological processes that make up the long, complicated carcinogenic process, it seems highly probable that physiological-adaptive and confrontational-adversarial components are both present at different times and to different degrees. It appears attractive to me to consider the early and intermediate steps in the carcinogenic process as mainly physiological-adaptive, while the stages of frank malignancy with progression as showing more confrontational-adversarial properties.

The very early initial interactions of mutagenic chemical carcinogens, radiations, and viruses with DNA would obviously prejudice most of us to consider the adversarial "abnormal" view as the appropriate one. Yet, I cannot overemphasize the unusually common nature of the earliest altered rare cells that appear during carcinogenesis, their unusually bland nature, and their spontaneous differentiation to normal-appearing adult liver. In my opinion, there is virtually no evidence to support the view that the rare altered cells appearing after initiation are in any way "abnormal" *in their behavior*. The finding of structural alterations in some genes after initiation does not prove by any stretch of the imagination that they are playing any role as determinants and that the new rare cell is biologically a mutant.

REFERENCES

1. Farber, E., and R. Cameron. "The Sequential Analysis of Cancer Development," *Adv. Cancer Res.* 31:125–226 (1980).
2. Farber, E. "Pre-Cancerous Steps in Carcinogenesis in Their Physiological Adaptive Nature," *Biochim. Biophys. Acta* 738:171–180 (1984).
3. Farber, E., and D. S. R. Sarma. "Biology of Disease: Hepatocarcinogenesis—A Dynamic Cellular Perspective," *Lab. Invest.* 56:4–22 (1987).
4. Farber, E. "Clonal Adaptation During Carcinogenesis," *Biochem. Pharmacol.* 39:1837–1846 (1990).
5. Farber, E., and H. Rubin. "Cellular Adaptation in the Development of Cancer," *Cancer Res.* 51:2751–2761 (1991).
6. Roomi, M. W., R. K. Ho, D. S. R. Sarma, and E. Farber. "A Common Biochemical Pattern in Preneoplastic Hepatocyte Nodules Generated in Four Different Models in the Rat," *Cancer Res.* 45:564–571 (1985).
7. Roomi, M. W., A. Columbano, G. M. Ledda-Columbano, and D. S. R. Sarma. "Lead Nitrate Induces Biochemical Properties Characteristic of Hepatocyte Nodules," *Carcinogenesis* 7:1643–1646 (1986).
8. Burt, R. K., S. Garfield, K. Johnson, and S. S. Thorgeirsson. "Transformation of Rat Liver Epithelial Cells with v-H-*ras* or v-*raf* Causes Expression of MDR-1, Glutathione S-Transferase P an Increased Resistance to Cytotoxic Chemicals," *Carcinogenesis* 9:2329–2332 (1988).
9. Lee, L. W., M.-S. Tsao, J. W. Grisham, and G. J. Smith. "Emergence of Neoplastic Transformants Spontaneously or after Exposure to N-Methyl-N'-nitro-N-nitrosoguanidine in Populations of Rat Liver Epithelial Cells Cultured

under Selective and Non-Selective Conditions," *Am. J. Pathol.* 135:63–71 (1989).

10. Tatematsu, M., Y. Nagamine, and E. Farber. "Redifferentiation as a Basis for Remodeling of Carcinogen-Induced Hepatocyte Nodules to Normal Appearing Liver," *Cancer Res.* 43:5049–5058 (1983).

11. Farber, E., S. Parker, and M. Gruenstein. "The Resistance of Putative Premalignant Liver Cell Populations, Hyperplastic Nodules, to the Acute Cytotoxic Effects of Some Carcinogens," *Cancer Res.* 36:3879–3887 (1976).

12. Judah, D. J., R. F. Legg, and G. E. Neal. "Development of Resistance to Cytotoxicity During Aflatoxin Carcinogenesis," *Nature* 265:343–345 (1977).

13. Rinaudo, J. A. S., and E. Farber. "The Pattern of Metabolism of 2-Acetylaminofluorene in Carcinogen-Induced Hepatocyte Nodules in Comparison to Normal Liver," *Carcinogenesis* 7:523–528 (1986).

14. Huitfeldt, H. S., E. F. Spangler, J. M. Hunt, and M. C. Poirier. "Immunohistochemical Localization of DNA Adducts in Rat Liver Tissue and Phenotypically Altered Foci During Oral Administration of 2-Acetylaminofluorene," *Carcinogenesis* 7:123–129 (1986).

15. Harris, L., L. E. Morris, and E. Farber. "The Protective Value of a Liver Initiation-Promotion Regimen against the Lethal Effect of Carbon Tetrachloride in the Rat," *Lab. Invest.* 61:467–470 (1989).

16. Gupta, R. C., K. Earley, and F. F. Becker. "Analysis of DNA Adducts in Putative Premalignant Hepatic Nodules and Nontarget Tissues of Rats during 2-Acetylaminofluorene Carcinogenesis," *Cancer Res.* 48:5270–5274 (1988).

17. Huitfeldt, H. S., P. Brandtzaeg, and M. C. Poirier. "Reduced Aminofluorene-DNA Adduct Formation in Replicating Liver Cells During Continuous Feeding of a Chemical Carcinogen," *Proc. Nat. Acad. Sci.* 87:5955–5958 (1990).

18. Enomoto, K., and E. Farber. "Kinetics of Phenotypic Maturation of Remodeling of Hyperplastic Nodules During Liver Carcinogenesis," *Cancer Res.* 42:2330–2335 (1982).

19. Rotstein, J., D. S. R. Sarma, and E. Farber. "Sequential Alterations in Growth Control and Cell Dynamics of Rat Hepatocytes in Early Precancerous Steps in Hepatocarcinogenesis," *Cancer Res.* 46:2377–2385 (1986).

20. Farber, E. "Hepatocyte Proliferation in the Stepwise Development of Experimental Liver Cell Cancer," *Digest. Dis. Sci.* 36:973–978 (1991).

21. Schulte-Hermann, R. "Induction of Liver Growth by Xenobiotic Compounds and Other Stimuli," *CRC Crit. Rev. Toxicol.* 3:97–158 (1974).

22. Schulte-Hermann, R. "Reactions of the Liver to Injury. Adaptation." in *Toxic Injury of the Liver,* E. Farber and M. M. Fisher, Eds. (New York: Marcel Dekker, 1979), pp. 385–444.

23. Stutman, O. "Immunodepression and Malignancy," *Adv. Cancer Res.* 22:261–422 (1975).

CHAPTER 7

Biostatistical Approaches for Modeling U-Shaped Dose-Response Curves and Study Design Considerations in Assessing the Biological Effects of Low Doses

Tom Downs, School of Public Health, University of Texas, Health Science Center, Houston, Texas

The first part of this two-part chapter deals with the probabilities of determining qualitatively what kinds of health effects may result from exposures to substances, and the second part with characterizing quantitative relationships between such health effects and exposures. The health effects may be beneficial in some situations, and detrimental in others.

QUALITATIVE RISK-BENEFIT EVALUATION

Introduction

Results of a study are customarily called "positive" if they show a positive association between substance exposure and disease, and "negative" otherwise.[1] Negative results thus include those with beneficial effects as well as those with no effect. But there have been numerous reports of substances showing beneficial effects at low doses and detrimental effects at high doses,[2,3] so the customary terminology could be confusing. To avoid ambiguities, results are referred to herein as harmful, neutral, or beneficial, according to whether there is a statistically significant harmful or high effect, no significant difference, or a significant low or beneficial effect.

The classification of a study result as one of these three may be incorrect because of sampling or other sources of variability. Of paramount importance in the design and planning of studies is the choice of study conditions that maximize, so far as is feasible, the probability of correctly classifying exposure effects as harmful, neutral, or beneficial. These probabilities depend on parameters such as sample size, dose regimen, duration of the

Table 7.1. Possible Results for a Study

	Study Result		
Actual Effect	Harmful	Neutral	Beneficial
Harmful	TH	FN	FL
Neutral	FH	TN	FL
Beneficial	FH	FN	TL
Unspecified	H	N	L

study, measure of outcome employed, spontaneous rate of the disease, and the number and importance of other potential confounding variables that may be present.[4] Some of these that are under the investigators' control will be examined for the effects that their manipulation prior to beginning a study may have on the probability of correctly classifying a study outcome.

For any study the possible results may be arrayed as in Table 7.1. *H*, *N*, and *L* denote harmful, neutral, and beneficial, respectively, and *T* and *F* denote true and false. Thus, if a study of a substance whose effects are actually beneficial results in a statistically significant harmful result, the study result is a *FH* (false high); if the actual effect is unspecified, then the result is simply *H*. Whatever the actual effect of a substance, the result of the study will be either *H*, *N*, or *L*. The probabilities of these three events thus sum to unity for each of the four scenarios in the rows of the table.

In testing for beneficial effects, it is essential that there be a high spontaneous rate in the unexposed group, so the beneficial effect can manifest itself. For if the disease outcome were extremely rare, then even a zero disease rate in the exposed subjects would not be unusual, and the sensitivity of the study to detect beneficial effects would be greatly impaired. In such cases either more sensitive subjects might be found, or case-control studies might be employed since these may assure sufficient numbers of disease cases. Failing these or other techniques to increase sensitivity to detect a TL, serious consideration should be given to abandoning the study.

Data Layout: Two-by-Two Tables

In many studies the relevant study data can be put in the form of a two-by-two table, as shown by Table 7.2. The study subjects are presumed to have been randomly allocated to the exposure groups: n_1 in the exposed group and n_0 in the nonexposed comparison group. These are then exposed

Table 7.2. Sample Two-by-Two Table

	Disease		
Exposure	Yes	No	Total
Yes	a	b	n_1
No	c	d	n_0
Total	m_1	m_0	n

or not, accordingly, and followed in time to observe ultimately the numbers of subjects who acquire the specific disease in question.

These numbers, a and c, in the exposed group and in the comparison group, will be independent in well-conducted studies and will be binomially distributed as, say:

$$a \sim B(n_1, p_1), \quad c \sim B(n_0, p_0)$$

where p_1 and p_0 are the disease rates in the exposed and comparison groups, respectively. Since the sample sizes n_1 and n_0 have been fixed in advance, knowledge of the numbers of diseased cases, a and c, will be sufficient to determine the numerical values of all numbers in the table.

Exposure to the test substance will have a beneficial effect with regard to the disease in question, and subject to the conditions of the study, when p_1 is less than p_0; no effect when p_1 equals p_0; and a harmful effect when p_1 is higher than p_0. A U-shaped hormetic dose-response exists when p_1 is less than p_0 at lower doses, and higher than p_0 at higher doses.

We shall examine how changes in the sample sizes n_1 and n_0 and in the disease rates p_1 and p_0 affect the probabilities that the study outcome will be significantly beneficial or harmful. Regardless of the actual effect of the test substance, these parameters have a strong influence on the probabilities. It is possible, through various means, to exert some control over these. Thus, the spontaneous rate p_0 in the comparison group may be changed by choosing more or less susceptible subjects, by varying the duration of follow-up for the study, by altering the environment in which the study takes place, by limiting or increasing the intensity of case-finding, by modifying the diet, etc. The rate p_1 for the exposed group may be altered by these same devices, and if the substance does indeed have an effect, then p_1 may also vary with such things as route of administration, dose regimen, and synergistic effects with existing environmental conditions.

The dose regimen is crucially important to the study of hormetic effects. It has been customary to employ relatively high doses in toxicological research, and then to extrapolate detrimental effects downward to lower dose levels. Low doses have seldom been employed. Such customs guarantee failure to detect any genuine beneficial effects at low doses.

Construction of High, Neutral, and Low Regions

It will be assumed the data in the two-by-two table have been analyzed via a two-sided chi-square test using Yates' continuity correction. Values of chi-square greater than 2.706 are statistically significant at the 10% significance level with a two-sided test (5% in each tail), and provide evidence of a beneficial effect when the odds ratio ad/bc is lower than unity, a neutral effect when the odds ratio is equal to unity, and a detrimental effect when it is higher than unity. Values of chi-square greater than 3.841 are interpreted similarly, but at the 5% level (with 2.5% in each tail).

Assume the number of subjects in each group is 25: $n_0 = n_1 = 25$. Setting the expression for chi-square equal to 2.706:

$$n(|ad-bc|-n/2)^2/(m_1 m_0 n_1 n_0) = 2.706$$

determines, for any value of c, those integral values of a for which the study result (a,c) will just be statistically significantly high or low, given the assumption that the null hypothesis, $p_1 = p_0$, is true.

Assigning a value of zero to c in the equation yields two quadratic equations in a — one with the odds ratio lower than unity, indicating some beneficial effect, and the other with the odds ratio exceeding unity, a harmful effect. The first equation has only imaginary solutions, indicating that there is no possible value for a that could significantly improve on an observed rate of zero in the nonexposed group. The second equation has the admissible solution a = 5, indicating that 5 or more diseased subjects in the exposed group would be significantly higher than the zero number in the comparison group.

This process is repeated with c = 1, yielding an imaginary solution again in the first case, and the admissible solution a = 6 in the second. The process is repeated again with c = 2, then again with c = 3, and c = 4. In all these instances the spontaneous rate is too low, and no L result is possible. At c = 5 though, a value of zero for a would just yield a significant low result. This process continues until finally c = 20 is reached. Here it is found that the minimum value of a required to achieve a significant H result is 25, the entire exposed group. When c is 21 or more there is no admissible value for a that will provide a significant H result.

The results for all possible combinations of a and c may be viewed graphically by constructing a 26-by-26 grid, with c on the horizontal axis and a on the vertical. By the process just described, this ac square is partitioned into three regions, H, N, and L, in Figure 7.1, according to the nature of the result. It should be kept in mind that these regions, H, N, and L, have been constructed under the assumption that the disease rates are the same in both groups. This assumption establishes the distribution for chi-square. The N region is concentrated around the diagonal of the square, where a = c. The H region has a larger than c, and the L region has a smaller than c. By the manner in which the significant values for a were obtained, the probability that a lies in the H region is not more than 5% for any value of c. This probability is not exactly 5% because of the discrete nature of the observations. Indeed, in many cases the probability is considerably smaller than 5%. Thus the "target" 5% is only a nominal figure. Similar remarks apply to the L region. In contrast, the probability of a result (a,c) falling in the N region is at least 90% when the null hypothesis is true.

From Figure 7.1, it is not possible to obtain an L result when there are fewer than 5 disease cases in the unexposed group. Thus, studies aimed at finding beneficial effects with only 25 subjects per group have very poor

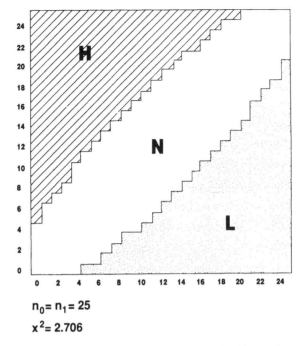

$n_0 = n_1 = 25$

$x^2 = 2.706$

Figure 7.1.   The H, N and L regions of possible study results. Exposed cases are on the vertical axis and non-exposed cases on the horizontal axis.

sensitivity unless the spontaneous rate exceeds 20%. Likewise, studies with more than 20 cases in the unexposed group have no chance at all of demonstrating an H effect. Even when there are 50 subjects in each group, though, there must still be at least 5 diseased subjects in the unexposed group before a significantly beneficial effect can occur. But now the 5 cases represent only 10% of the sample of 50, whereas before the 5 cases represented 20% of the sample.

Probabilities of Study Outcomes

For any given sample sizes n_1 and n_0, not necessarily equal, an ac rectangle can be constructed as above. The calculations involved are entirely independent of the actual values of the rates p_1 and p_0, and depend only on the assumed null hypothesis that they are equal. The H, N, and L regions define the result for any possible outcome of the study because, for fixed sample sizes n_1 and n_0, the study outcome is completely specified by the values for a and c.

As stressed above, the 10% significance figure is only nominal. But, while the H, N, and L regions were constructed using this 10% for defining the value of chi-square, the actual probabilities of a study result, (a,c), falling in any particular region, including the null N region, depend strongly on the numerical values of p_0 and p_1, even when these are equal. For example, the probability of an H or L result is zero when the study conditions impose a zero disease rate for both groups. It is also zero when the disease is universally prevalent in both groups.

Thus, in order to calculate the probability of a study outcome (a,c) landing in the H, N, or L zones, it is necessary to specify the rates p_1 and p_0 as well as the corresponding sample sizes n_1 and n_0. Since a and c are independent and binomially distributed, the probability of any particular pair (a,c) in the ac rectangle is the product of the corresponding binomial probabilities:

$$\Pr(a)\Pr(c) = \frac{n_1! \, n_0!}{a!s!c!t!} \; p_1^a \, (1 - p_1)^s \; p_0^c \, (1 - p_0)^t$$

where $s = n_1 - a$ and $t = n_0 - c$. The probability that a study outcome pair falls in the L region, say, is then the sum of all such probabilities over all pairs (a,c) in the L region.

The probabilities of a study result falling in the L region have been calculated for a variety of scenarios and appear in Table 7.3 (2.5% in each tail) and Table 7.4 (5% in each tail), rounded to the nearest percent. The probabilities along the diagonals correspond to L study results when, in fact, the actual effect of the substance is N (so such a result is a FL). These probabilities, all less than 2.5% by design, vary in Table 7.3A from 0% when $p_1 = p_0 = 0.05$ to 2% when $p_1 = p_0 = 0.50$. The probabilities above the diagonal, all much smaller still, correspond to L study results when, in fact, the substance is harmful (FL again). The probabilities below the diagonal correspond to L study results when, indeed, the substance is beneficial (TL).

The percents below the diagonal are thus the sensitivities for detecting a true beneficial effect. They increase as the difference $p_0 - p_1$ increases, and for fixed $p_0 - p_1$ tend to increase as p_1 and p_0 both decrease. The tables show the probability of a true beneficial result increases with sample size. The diagonal entries in the tables tend to approach their nominal values of 2.5 and 5% as the sample size increases, because the larger samples lessen the effect of the discreteness of the observations. For a fixed total sample size, maximum sensitivities for detecting true H or L effects are achieved when the sample sizes n_1 and n_0 are equal.

All the above analyses have been done under the assumption that the sample sizes n_1 and n_0 for the exposed and nonexposed groups were determined in advance of the actual study. The analyses are actually applicable to a much wider class of studies and hold when the sample sizes m_1 and m_0 for disease cases and noncases are determined in advance, instead of the numbers of exposed and unexposed subjects, because the chi-square analysis is appropriate for both cases. Instead of ac rectangles for cases there are now ab rectangles for exposed subjects. And instead of the null hypothesis being equal disease rates it is now equal exposure rates. All this is simply accomplished in practice by interchanging the exposure-disease labeling for the rows and columns in the data table while leaving everything else intact; then all the original notation and results are applicable.

Data Layout: Standardized Ratios

In many nonexperimental epidemiological studies, the study subjects form a well-defined cohort, such as all the employees, past and present, of a petrochemical plant. In such studies the outcome measure often takes the form of a *standardized ratio* (SR). The SR is defined as the ratio of the observed number of disease events occurring among the members of the study population, to some theoretically derived number. This theoretical number is the number of cases that would ordinarily be expected to occur if the study population suffered disease rates similar to a "standard" population whose rates were used to obtain the expected number.

Briefly, the expected numbers are calculated by applying the standard population rates—usually on a race-, age-, sex-, time-, place-, and disease-specific basis—to the study population over the time frame of the study. The resulting SR ratio of observed-to-expected cases is called a *standardized mortality ratio* (SMR) if the outcome is death from a specific cause, and a *standardized incidence ratio* (SIR) if the study outcome is contraction of a specific disease.

Most studies of the effects of occupational exposures on worker health employ some form of standardized ratio to measure the outcome. The study results are usually reported separately by race, sex, and disease. Results for a wide variety of diseases and subgroups of the study population are ordinarily reported.[5] Computation of the expected numbers becomes cumbersome when the data sets are large. A vast array of statistical rate tables is also required. Computer programs with built-in rate tables have been written and employed to carry out these mechanical tasks.[6] The program developed by Monson ordinarily computes 40–50 standardized ratios for each subgroup of workers. These subgroups may be cross-classified (e.g., by race, sex, exposure histories, length of employment), so the number of SRs generated can become very large.

When the SR is lower than unity, the observed number of events is less than the expected number, and a beneficial effect has been observed. When the SR is unity, the effect is neutral, and when the SR is higher than unity, the observed number is greater than the expected, and a detrimental effect occurs. In terms of statistical significance (e.g., false high results, etc.), the same terminology is used here as above, with H and L representing statistically significant SRs, and N a neutral SR that is not significantly different from unity.

There are two serious sources of bias commonly affecting SRs. First is the bias toward significant L effects due to the "healthy worker effect"—

Table 7.3. Probability of Observing a Significantly Lower Risk in the Exposed Group, by Sample Size, Risk for Exposed Group, and Risk for Comparison Group

Risk for Exposed Group	Risk for Comparison Group									
	0.05	0.10	0.15	0.20	0.25	0.30	0.35	0.40	0.45	0.50
A. Sample Size = 25 in Each Group										
0.50	0	0	0	0	0	0	0	0	1	2
0.45	0	0	0	0	0	0	0	1	2	4
0.40	0	0	0	0	0	0	1	1	4	8
0.35	0	0	0	0	0	0	1	3	7	14
0.30	0	0	0	0	0	1	3	7	13	23
0.25	0	0	0	0	1	3	6	13	23	36
0.20	0	0	0	1	3	6	13	23	36	51
0.15	0	0	1	2	6	14	24	38	53	67
0.10	0	0	2	6	15	27	42	57	72	83
0.05	0	1	6	16	31	49	66	79	89	95
B. Sample Size = 50 in Each Group										
0.50	0	0	0	0	0	0	0	0	0	2
0.45	0	0	0	0	0	0	0	0	2	5
0.40	0	0	0	0	0	0	0	2	5	13
0.35	0	0	0	0	0	0	2	5	13	27
0.30	0	0	0	0	0	2	5	13	29	46
0.25	0	0	0	0	2	5	14	29	48	67
0.20	0	0	0	1	6	15	32	52	71	85
0.15	0	0	1	6	17	35	57	76	89	96
0.10	0	1	6	19	41	64	82	93	98	99
0.05	0	6	24	51	75	90	97	99	100	100

Table 7.3 cont.

C. Sample Size = 75 in Each Group

0.50	0	0	0	0	0	0	0	0	2
0.45	0	0	0	0	0	0	0	2	7
0.40	0	0	0	0	0	0	2	7	20
0.35	0	0	0	0	0	2	7	20	42
0.30	0	0	0	0	2	7	20	42	67
0.25	0	0	0	2	7	21	43	68	86
0.20	0	0	1	8	23	47	71	88	97
0.15	0	1	8	26	53	77	92	98	100
0.10	1	10	32	61	84	95	99	100	100
0.05	13	44	76	93	98	100	100	100	100

D. Sample Size = 100 in Each Group

0.50	0	0	0	0	0	0	0	0	2
0.45	0	0	0	0	0	0	0	2	9
0.40	0	0	0	0	0	0	2	9	26
0.35	0	0	0	0	0	2	8	26	53
0.30	0	0	0	0	2	9	26	53	79
0.25	0	0	0	2	9	29	56	81	94
0.20	0	0	2	10	31	61	84	96	99
0.15	0	1	11	36	67	89	97	100	100
0.10	1	13	42	75	93	99	100	100	100
0.05	17	57	87	98	100	100	100	100	100

Note: Test statistic is continuity-corrected chi-square with 1 df. Two-tailed test with nominal probability of 2.5% in each tail.

Table 7.4. Probability of Observing a Significantly Lower Risk in the Exposed Group, by Sample Size, Risk for Exposed Group, and Risk for Comparison Group

Risk for Exposed Group	Risk for Comparison Group									
	0.05	0.10	0.15	0.20	0.25	0.30	0.35	0.40	0.45	0.50
A. Sample Size = 25 in Each Group										
0.50	0	0	0	0	0	0	0	1	1	3
0.45	0	0	0	0	0	0	0	1	3	7
0.40	0	0	0	0	0	0	1	3	7	13
0.35	0	0	0	0	0	1	3	6	12	21
0.30	0	0	0	0	1	3	6	12	21	33
0.25	0	0	0	1	3	6	12	21	34	48
0.20	0	0	1	2	6	13	23	35	49	64
0.15	0	0	2	6	13	24	38	53	67	79
0.10	0	1	5	13	26	42	58	72	83	91
0.05	0	3	12	27	46	64	79	89	95	98
B. Sample Size = 50 in Each Group										
0.50	0	0	0	0	0	0	0	0	1	3
0.45	0	0	0	0	0	0	0	1	3	8
0.40	0	0	0	0	0	0	1	3	8	19
0.35	0	0	0	0	0	1	3	9	20	36
0.30	0	0	0	0	1	3	10	21	38	57
0.25	0	0	0	1	3	9	22	41	61	78
0.20	0	0	1	3	10	23	42	64	81	92
0.15	0	0	3	11	26	46	67	84	94	98
0.10	0	2	12	31	54	75	89	96	99	100
0.05	1	13	39	66	85	95	99	100	100	100

Table 7.4 cont.

C. Sample Size = 75 in Each Group

0.50	0	0	0	0	0	0	0	0	1	3
0.45	0	0	0	0	0	0	0	1	3	11
0.40	0	0	0	0	0	0	1	4	12	28
0.35	0	0	0	0	0	1	4	13	30	53
0.30	0	0	0	0	1	3	12	30	55	77
0.25	0	0	0	1	3	13	31	56	79	92
0.20	0	0	0	3	14	34	59	81	93	98
0.15	0	0	3	15	38	65	85	95	99	100
0.10	0	3	16	44	73	91	98	100	100	100
0.05	2	19	55	84	96	99	100	100	100	100

D. Sample Size = 100 in Each Group

0.50	0	0	0	0	0	0	0	0	1	4
0.45	0	0	0	0	0	0	0	1	4	14
0.40	0	0	0	0	0	0	1	4	14	36
0.35	0	0	0	0	0	1	4	14	36	64
0.30	0	0	0	0	1	4	15	37	65	87
0.25	0	0	0	1	3	15	40	69	88	97
0.20	0	0	0	3	17	43	72	91	98	100
0.15	0	0	3	19	48	77	93	99	100	100
0.10	0	3	22	57	85	97	100	100	100	100
0.05	2	28	71	94	99	100	100	100	100	100

Note: Test statistic is continuity-corrected chi-square with 1 df. Two-tailed test with nominal probability of 5% in each tail.

workers are self-selected because they were healthy enough to work, and thus they may form a biased group in relation to the "standard" population they are being compared to. This potential bias could, in theory, be avoided by choosing the standard population more appropriately; but in practice this is not possible because the rates for calculating expected numbers are ordinarily only compiled for the population as a whole – and not special subgroups like the working population. Thus, significant beneficial effects observed in an occupational cohort should be viewed cautiously; they may be real, or they may be manifestations of the healthy worker effect. The situation is further exacerbated by the fact that the magnitude of the healthy worker effect varies from one disease to another; it is likely to be most pronounced in those diseases that occur and are debilitating early in life, and less pronounced in those diseases, such as cancer, that occur later in life.[7,8] The healthy worker effect is most pronounced in younger workers and diminishes over time as the workers age.

The second source of bias has historically been in the opposite direction, toward false high results. These are brought about by wholesale computation of vast combinations, arrays, and varieties of tests of statistical hypotheses with the aim of finding statistically significant results. This phenomenon has been dubbed the "multiple comparison problem" by statisticians. The proliferation of computers and statistical software in recent years has aggravated the situation.

Construction of High, Neutral, and Low Regions

There is a wealth of sound theoretical and empirical evidence to the effect that, as a general rule, the number of observed cases or deaths in a defined group of subjects follows a Poisson probability distribution. Thus, this distribution may then be used to construct H, N, and L regions for testing the null hypothesis that the observed number has average value equal to the expected number, as calculated above from standard rates.

Construction of such regions is much simpler than for two-by-two tables since the Poisson distribution (and hence the regions) are determined completely by just the expected number of cases. Bailar and Ederer provide tables for two-sided tests of significance of the null hypothesis that the SR is unity.[9] Table 7.5 can be used to obtain 95% "normal limits" for the observed number of cases as a function of the expected number of cases. It can also be used to construct a 95% confidence interval for the true Poisson mean as a function of the observed number of cases. In addition, it can be used to construct a 95% confidence interval for the true SR as a function of the observed and the expected number of cases.

To find normal limits for the observed number of cases as a function of the expected number, locate the smallest entry in Table 7.5A that is as big or bigger than the expected number. Then

lower normal limit = (10 × row number) + (column number)

The upper normal limit is obtained similarly from Table 7.5B:

upper normal limit = (10 × row number) + (column number)

For example, if the expected number of cases is 15.21, then the observed number of cases should be between 8 and 23, inclusive, at least 95% of the time. Observed values below 8 or above 23 constitute significant L or H results, respectively.

To find 95% confidence limits for the true expected number as a function of the observed number, convert the observed number to a row and column via the formula

(observed number) = (10 × row number) + (column number)

Then the confidence limits for the true expected number are

lower limit = Table 7.5B entry for (observed number – 1)

upper limit = Table 7.5A entry for (observed number)

If (observed number) of cases is 30, then (observed number − 1) is 29, so that

lower limit = Table 7.5B entry for (29) = 20.24

upper limit = Table 7.5A entry for (30) = 42.83

To get a confidence interval for an SR, suppose the observed number of cases is 30 and the expected number is 15.21. Then the SR is observed/expected = 30/15.21, nearly double, and the confidence interval (CI) for the SR is

$$95\% \text{ CI for SR} = (95\% \text{ CI for true expected})/(\text{expected})$$
$$= (20.24, 42.83)/15.21$$
$$= (1.33, 2.82)$$

Since the lower limit of the CI is greater than one, the result is H. Alternatively, the observed 30 cases exceeds the upper normal limit of 23, so the result is H. As a third approach, the hypothesized expected number of 15.21 is lower than the lower confidence limit of 20.24, so again the result is H. These three methods give the same result, but emphasize different aspects.

Probabilities of Study Outcomes

Beaumont and Breslow discuss power calculations in occupational settings when the Poisson distribution is involved.[10] As a general rule, the larger the expected number, the greater the power to detect H or L effects. When the expected number is less than 3.69, then even an observed value of

Table 7.5. Poisson Distribution 95% Normal Limits for Observed Number of Events, 95% Confidence Limits for Expected Number of Events, and 95% Confidence Limits for Standardized Ratios of Observed to Expected Number of Events

A. Lower Normal Limit or Upper Confidence Limit

	0	1	2	3	4	5	6	7	8	9
0	3.60	5.57	7.22	8.77	10.24	11.67	13.06	14.42	15.76	17.08
1	18.39	19.68	20.96	22.23	23.49	24.74	25.98	27.22	28.45	29.67
2	30.89	32.10	33.31	34.51	35.71	36.90	38.10	39.28	40.47	41.65
3	42.83	44.00	45.17	46.34	47.51	48.68	49.84	51.00	52.16	53.31
4	54.47	55.62	56.77	57.92	59.07	60.21	61.36	62.50	63.64	64.78
5	65.92	67.06	68.19	69.33	70.46	71.59	72.72	73.85	74.98	76.11
6	77.23	78.36	79.48	80.61	81.73	82.85	83.97	85.09	86.21	87.32
7	88.44	89.56	90.67	91.79	92.90	94.01	95.12	96.24	97.35	98.46
8	99.57	100.67	101.79	102.89	104.00	105.10	106.21	107.31	108.42	109.52
9	110.63	111.73	112.83	113.93	115.03	116.13	117.23	118.33	119.43	120.53
10	121.63	122.72	123.82	124.92	126.01	127.11	128.20	129.30	130.39	131.48
11	132.58	133.67	134.76	135.85	136.95	138.04	139.13	140.22	141.31	142.40
12	143.49	144.58	145.67	146.76	147.85	148.93	150.02	151.10	152.19	153.28
13	154.36	155.45	156.53	157.62	158.71	159.79	160.87	161.96	163.04	164.12
14	165.20	166.29	167.37	168.45	169.54	170.62	171.70	172.77	173.86	174.94
15	176.01	177.10	178.18	179.25	180.34	181.41	182.49	183.57	184.64	185.72
16	186.80	187.88	188.96	190.04	191.11	192.19	193.27	194.34	195.42	196.49
17	197.56	198.64	199.72	200.79	201.86	202.94	204.01	205.08	206.16	207.23
18	208.30	209.37	210.45	211.52	212.59	213.67	214.73	215.81	216.88	217.96
19	219.03	220.09	221.17	222.24	223.30	224.37	225.45	226.52	227.59	228.65
20	229.72	230.79	231.86	232.92	234.00	235.07	236.14	237.20	238.27	239.34

Table 7.5 cont.

B. Upper Normal Limit or Lower Confidence Limit

	0	1	2	3	4	5	6	7	8	9
0	0.03	0.24	0.62	1.09	1.62	2.20	2.81	3.45	4.12	4.80
1	5.49	6.20	6.92	7.65	8.40	9.15	9.90	10.67	11.44	12.22
2	13.00	13.79	14.58	15.38	16.18	16.98	17.79	18.61	19.42	20.24
3	21.06	21.89	22.72	23.55	24.38	25.21	26.05	26.89	27.73	28.58
4	29.42	30.27	31.12	31.97	32.82	33.68	34.53	35.39	36.25	37.11
5	37.97	38.84	39.70	40.57	41.43	42.30	43.17	44.04	44.91	45.79
6	46.66	47.54	48.41	49.29	50.17	51.04	51.92	52.80	53.69	54.57
7	55.45	56.34	57.22	58.11	58.99	59.88	60.77	61.66	62.55	63.44
8	64.33	65.22	66.11	67.00	67.89	68.79	69.68	70.58	71.47	72.37
9	73.27	74.17	75.06	75.96	76.86	77.76	78.66	79.56	80.46	81.36
10	82.27	83.17	84.07	84.98	85.88	86.78	87.69	88.59	89.50	90.41
11	91.31	92.22	93.13	94.04	94.94	95.85	96.76	97.67	98.58	99.49
12	100.40	101.31	102.23	103.14	104.05	104.96	105.87	106.79	107.70	108.61
13	109.53	110.44	111.36	112.27	113.19	114.10	115.02	115.94	116.85	117.77
14	118.69	119.61	120.52	121.44	122.36	123.28	124.20	125.12	126.04	126.96
15	127.88	128.80	129.72	130.64	131.56	132.48	133.40	134.32	135.25	136.17
16	137.09	138.01	138.94	139.86	140.78	141.71	142.63	143.56	144.48	145.41
17	146.33	147.26	148.18	149.11	150.03	150.96	151.88	152.81	153.74	154.66
18	155.59	156.52	157.45	158.37	159.30	160.23	161.16	162.09	163.01	163.94
19	164.87	165.80	166.73	167.66	168.59	169.52	170.45	171.38	172.31	173.24
20	174.17	175.10	176.03	176.96	177.90	178.83	179.76	180.69	181.62	182.56

Note: See text for explanation of table.

zero, the lowest value possible, is not unusual since zero is the lower normal limit of a 95% normal range (Table 7.5A).

Tables 7.6 and 7.7 contain sensitivity probabilities for detecting hormetic effects in SR studies. These depend on the relative risk (RR), defined in the present context by

$$RR = \text{(true expected number)/(calculated expected number)}$$

If the null hypothesis of no difference in risk is true, then the RR is unity. In such case (the last columns in Tables 7.6 and 7.7), the actual effect is neutral, and the corresponding tabular entries are the probabilities of Type I errors. These cannot exceed 2.5%, and are often considerably below 2.5% because of the discreteness of the Poisson distribution. For all the other columns, where the RR is lower than one, there is a hypothesized beneficial effect relative to the standard population, and the tabular entries are the probabilities of TLs for the corresponding hypothetical RRs. These sensitivity probabilities tend to increase as the expected cases increase (though in an erratic manner due to discreteness) and do increase systematically as the RR decreases.

Meta-Analytic Techniques

Sometimes there may be several studies available with low-level exposures to a substance, yet each individual study has insufficient sensitivity to detect a hormetic effect. However, when the studies are pooled or viewed collectively, a clear pattern may emerge. Meta-analysis is a collection of statistical techniques for objectively synthesizing or pooling similar studies. Meta-analytic results are not always definitive. An attempt to synthesize the literature on effects of passive smoking via meta-analytic techniques yielded inconclusive results.[11]

There are numerous philosophical and practical problems involved in meta-analysis. Many of these stem from dissimilarities between the several studies being synthesized. No two studies are exact duplicates of each other, and it may be impossible to statistically adjust for differences between them in a satisfactory way. Meta-analysis has been described by some as controversial and by others as the wave of the future.[12] Like any statistical technique, it can be misused. A list of problems and questions concerning meta-analysis has been compiled by Spitzer.[13]

Meta-analytic studies of hormetic effects can be expected to suffer from "publication bias"—nondetrimental results tend not to be submitted for publication by investigators, and such nondetrimental results that are submitted tend to be rejected by editors. Publication bias could result in significant overestimates of risk in humans and be detrimental to rational health policy decisions. A group of epidemiologists recently concluded that publication bias was a definite problem in their field.[1,14] Nevertheless meta-

Table 7.6. Probability of Observing a Significantly Lower Risk in an Exposed Group, by Relative Risk and Expected Number of Cases in the Exposed Group When the Relative Risk is Unity

Expected Cases	Relative Risk									
	0.1	0.2	0.3	0.4	0.5	0.6	0.7	0.8	0.9	1.0
2	0	0	0	0	0	0	0	0	0	0
4	67	45	30	20	14	9	6	4	3	2
6	88	66	46	31	20	13	8	5	3	2
8	95	78	57	38	24	14	8	5	3	1
10	98	86	65	43	27	15	8	4	2	1
12	100	96	84	65	45	28	16	8	4	2
14	100	98	87	67	45	27	14	7	3	1
16	100	99	94	80	59	38	21	11	5	2
18	100	100	95	81	59	36	19	9	4	2
20	100	100	98	89	70	46	26	13	5	2
22	100	100	98	89	69	44	24	11	4	2
24	100	100	99	94	77	53	30	14	6	2
26	100	100	100	96	84	61	36	17	7	2
28	100	100	100	96	83	58	33	15	6	2
30	100	100	100	98	88	65	38	18	7	2
32	100	100	100	98	87	63	36	16	5	2
34	100	100	100	99	90	69	41	19	7	2
36	100	100	100	99	93	74	46	21	8	2
38	100	100	100	99	93	72	43	19	6	2
40	100	100	100	100	95	77	47	22	7	2
42	100	100	100	100	96	81	52	24	8	2
44	100	100	100	100	96	79	49	22	7	2
46	100	100	100	100	97	83	53	24	8	2
48	100	100	100	100	98	86	57	27	9	2
50	100	100	100	100	99	88	61	30	10	2

Note: Rounded to nearest percent. Test statistic is observed Poisson-distributed count of cases in exposed group. Nominal probability of 2.5% in lower tail.

Table 7.7. Probability of Observing a Significantly Lower Risk in an Exposed Group, by Relative Risk and Expected Number of Cases in the Exposed Group When the Relative Risk is Unity

Expected Cases	Relative Risk									
	0.1	0.2	0.3	0.4	0.5	0.6	0.7	0.8	0.9	1.0
2	0	0	0	0	0	0	0	0	0	0
4	67	45	30	20	14	9	6	4	3	2
6	88	66	46	31	20	13	8	5	3	2
8	99	92	78	60	43	29	19	12	7	4
10	100	95	82	63	44	29	17	10	5	3
12	100	99	93	79	61	42	27	16	9	5
14	100	99	94	80	60	40	24	13	7	3
16	100	100	97	89	72	51	32	18	9	4
18	100	100	98	89	71	48	29	15	7	3
20	100	100	99	94	79	58	36	19	9	4
22	100	100	100	96	85	65	43	24	11	5
24	100	100	100	96	84	63	39	20	9	3
26	100	100	100	98	89	70	45	24	11	4
28	100	100	100	99	92	75	51	28	13	5
30	100	100	100	99	92	73	47	24	10	4
32	100	100	100	99	94	78	52	28	12	4
34	100	100	100	100	96	82	57	31	13	5
36	100	100	100	100	96	80	54	28	11	3
38	100	100	100	100	97	84	58	31	12	4
40	100	100	100	100	98	87	62	34	14	4
42	100	100	100	100	98	89	66	37	15	5
44	100	100	100	100	98	88	63	33	13	4
46	100	100	100	100	99	90	67	36	14	4
48	100	100	100	100	99	92	70	39	15	4
50	100	100	100	100	99	94	73	42	17	5

Note: Rounded to nearest percent. Test statistic is observed Poisson-distributed count of cases in exposed group. Nominal probability of 5% in lower tail.

Table 7.8. Meta-Analysis Example for Saccharin and Bladder Neoplasia

Dose	FDA Study Hyperplasia (%)		WARF Study Tumors (%)		Japanese Study Tumors (%)
	Male	Female	Male	Female	Male
0	10/73 (14)	3/85 (4)	3/20 (15)	12/20 (60)	19/50 (38)
0.01	6/71 (8)	0/81 (0)	–	–	–
0.05	–	–	2/20 (10)	6/20 (30)	–
0.1	4/81 (5)	0/81 (0)	–	–	–
0.2	–	–	–	–	4/50 (5)
0.5	–	–	2/20 (10)	9/20 (45)	–
1.0	4/76 (5)	3/90 (3)	–	–	13/50 (26)
5.0	6/64 (9)	5/88 (6)	14/20 (70)	18/20 (90)	26/50 (52)
7.5	19/62 (31)	10/76 (13)	–	–	–

Note: FDA and WARF studies: incidence in rats; Japanese study: incidence in mice. Dose is percent of diet.

analysis can be a powerful tool. Despite publication bias meta-analysis may be the only way, in many instances, to demonstrate hormetic effects.

The data on saccharin and bladder neoplasia in Table 7.8 will illustrate the impact that multiple data sources may have. Sources for the data on saccharin and bladder neoplasia are provided by Downs and Frankowski.[15] For each of the five sex-species combinations the response declines at low doses in comparison to zero dose, plateaus, and then increases as the dose increases.

QUANTITATIVE DOSE-RESPONSE MODELS

Introduction

Almost without exception the dose-response models studied to date have focused on harmful effects. Current models thus have limited flexibility. Some contain mathematical restrictions prohibiting a decrease in response whenever there is an increase in dose. In such cases the existence of a threshold or of beneficial effects are excluded automatically from consideration. Some suggestions are proposed for development of more general models suitable for hormetic dose-response studies.

Model Criteria

The diversity of carcinogenic agents and responses, the variety of exposure settings and routes of administration, and the lack of detailed scientific knowledge about fundamental cellular processes in cancer all combine to make it unlikely that a single dose-response model can suffice for all situations in which a hormetic dose-response model might be required. Still, some general suggestions about models for U-shaped responses can be put forth:

1. The model should be flexible, adaptive, and parsimonious. If the parameters of the model are sufficiently versatile (e.g., capable of being scalar values or mathematical expressions of unspecified form), then these may be useful for looking at data in new ways and suggesting future avenues of research. Flexible models might be useful as guides for hypothesizing responses to low doses, thus aiding in designing dose regimens for hormetic studies, and in deciding whether a projected response is such that a proposed study will have sufficient statistical power. The flexibility requirement automatically rules out the one-hit linear model whose response P_x and slope dP_x/dx are given, for an administered dose x, by

$$P_x = 1 - \exp[-a - bx], \quad dP_x/dx = b \exp[-a - bx]$$

because the family of such curves is inflexible: All the curves in the family are convex upward at all doses for all positive values of the potency parameter b.

2. Above all, a model must be capable of exhibiting a U-shaped response. In particular, it must be possible for the slope dP_x/dx of the dose-response curve P_x to be negative for some small doses x, while eventually becoming positive for larger x. These are essential requirements. They automatically rule out virtually all the common models in use today. This includes the linearized multistage model, popular with regulatory agencies,[16] whose response has the form:

$$P_x = 1 - \exp[-q_0 - q_1 x - q_2 x^2 - \ldots]$$

since this model requires that the fitted coefficients $q_0, q_1, q_2, \ldots$ all be nonnegative.[16] Such nonnegativity forces the slope

$$dP_x/dx = (q_1 + 2q_2 x + \ldots) \exp[-q_0 - q_1 x - q_2 x^2 - \ldots]$$

to be positive for all non-zero doses x.

3. The hormetic components of a model must be scientifically verifiable. Models with a U-shape imposed on them by restricting parameters and mathematical forms should be avoided since implicit in such models is the assumption that there is a beneficial effect at all sufficiently low doses (i.e., there is no hormetic threshold). A quantitative measure of the hormetic effect should be attainable from the fitted model, and the hypothesized existence of a hormetic effect should be testable by statistical techniques.

4. Models should be capable of incorporating into their structure pertinent pharmacokinetic and biologic data. The parameters and mathematical forms of the model should be biologically interpretable.

5. Assumptions should be minimized. If made, they should be justified and checked whenever possible.

6. The suggestions above should not be unduly restrictive. The models ought to be applicable to a reasonably wide set of data, otherwise their usefulness is limited.

Adaptive Models

Adaptive Repair

Downs and Frankowski[15] employed Michaelis-Menten kinetics to develop a model with the potential for adaptive repair. For dose x, the probability R of repair has the form of a linear ratio:

$$R = \frac{px + q}{(p + r)x + (q + s)}$$

where the parameters p, q, r, and s are all nonnegative, insuring that R lies between 0 and 1. The number of "hits" from particles of the test substance or its metabolites on susceptible portions of DNA is assumed to be Poisson distributed, with mean the linear ratio

$$H = \frac{ax + b}{cx + d}$$

where here also all four coefficients are nonnegative. Note that when x is 0, the spontaneous repair rate is equal to q/(q + s), and the spontaneous hit rate is equal to b/d. The spontaneous rates of hits and repairs are here neither independent of nor additive with the rates induced by the test substance.

It was further assumed that the number of hits that were repaired followed a binomial distribution, with n equal to the number of hits and with the above R being the probability that any particular hit would be repaired. Then it is readily shown that the number of unrepaired hits follows a Poisson distribution, with mean equal to

$$y = H(1 - R).$$

Then y is a more valid measure of effective dose than the administered dose x. It is entirely possible that y can decrease as x increases, and in fact this will be the case whenever

$$\frac{dy}{dx} < 0$$

or

$$\frac{dR/dx}{1 - R} > \frac{dH/dx}{H} .$$

In such case the repair is adaptive, with the probability of repair increasing with increasing dose. Eventually though, H may become sufficiently large to overwhelm the enhanced repair (for a thorough discussion of these matters see Downs and Frankowski[15]).

The multistage model given by

$$P_x = 1 - \exp[-q_1 y - q_2 y^2 - . . .]$$

which uses y instead of x for the dose parameter, will exhibit hormetic behavior whenever $dy/dx < 0$ at small doses. Another model employing y instead of x would be the (now nonlinear) one-hit model

$$P_x = 1 - \exp[-y]$$

Graphs of this one-hit model are shown in Figure 7.2, where the hit rate H is given for each of the three curves by

$$H = \frac{10x + 10}{x + 10}$$

while the repair rate is given — for the top, middle, and bottom curves, respectively — by

$$R = \frac{85x + 9}{100x + 10}$$

$$R = \frac{95x + 9}{100x + 10}$$

$$R = \frac{99x + 9}{100x + 10}$$

Figure 7.2 illustrates the flexibility and adaptivity of this repair-modified one-hit model. In the top curve the repair rate is actually impeded by the test substance, and the response P_x increases dramatically at low doses. In the middle curve moderate doses are clearly beneficial, and in the bottom curve repair has been increased so much that the response P_x never gets back up to its spontaneous rate.

Any model, such as the k-hit, multistage, or probit,[17] can be modified to accommodate nonmonotonic dose-response data by using nonmonotonic functions of the dose, like y above, instead of the administered dose x.

Modified Linear Multistage Model

The multistage model given by

$$P_x = 1 - \exp[-q_0 - q_1 x - q_2 x^2 - . . .]$$

can be modified directly to accommodate the possibility of beneficial effects by merely allowing the linear coefficient q_1 to be negative. This would result in a beneficial effect whenever the dose x satisfies the inequal-

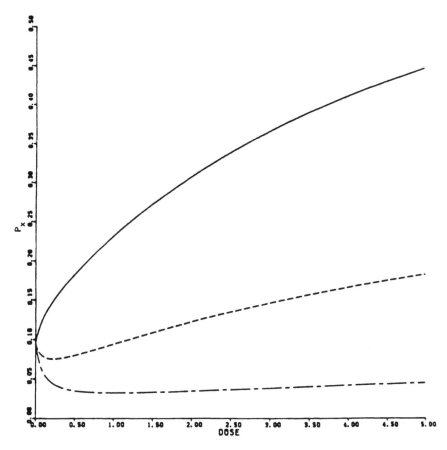

Figure 7.2.   Three dose-response curves illustrating the versatility of the one-hit model with adaptive repair.

ity $q_1x + q_2x^2 < 0$, or equivalently, $x < -q_1/q_2$. The optimal dose in this case is that value of x that minimizes P_x. It is obtained by setting the derivative of P_x equal to zero, solving the resulting equation for x, and verifying that the second derivative is positive when evaluated at this value of x.

The multistage model has the advantage of being characterized by a small number of parameters. These parameters may also be readily estimated from long-term animal bioassay data. This may also be a disadvantage, though, because it may be difficult to incorporate pharmacokinetic knowledge into the estimation process. Nevertheless, in the absence of additional biological data, the multistage model will often be the default choice in model-fitting because of its simplicity and parsimony.

Stochastic Two-Stage Model

Moolgavkar and Venzon developed a two-stage stochastic model for carcinogenesis, in which the stages may be thought of as mutations, though not necessarily so.[18] A schematic diagram of the model appears in Figure 7.3. Subsequently, Moolgavkar and Knudson expanded the scope of the model, showing how it could be interpreted biologically to explain or help explain such diverse phenomena as age-specific cancer rates, hereditary factors, and environmental agents in the etiology of cancer.[19]

When the mutation rate μ_2 of the second stage is small, the model implies that the age-specific incidence function I(t) for a tissue in a subject of age t is approximately given by

$$I(t) \sim \mu_1\mu_2 \int_0^t N(s)\exp[(\alpha_2-\beta_2)\ (t - s)]ds$$

where μ_1, μ_2, α_2, and β_2 are as in Figure 7.3, and N(s) is the number of normal stem cells in the tissue at time s. This approximation was employed by Moolgavkar[20] to give precise meanings to the words *initiator* and *promoter*. Later authors have expanded the scope of the model's usefulness in providing a framework for reviewing known facts.[21-23]

One of the strengths of this model is its versatility: The mathematical expressions for the mutation and birth and death rates of cells have been deliberately left unspecified, so that appropriate forms for such may be derived and tested to fit a variety of circumstances. Thus, while the model specifically requires both stages to be irreversible,[22] linear ratios incorporating cell repair notions, like R above, can still be employed and tested in the model. This requires some modification of the terminology: thus, μ_1 is now considered as the mutation rate for generating unrepaired intermediate cells from the normal stem cells. Hormetic effects are obtained from this model whenever the net change in cell growth is negative. This model, and varia-

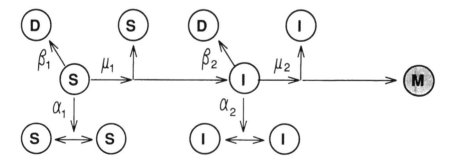

Figure 7.3. Two-stage model for carcinogenesis. S = normal stem cell, I = intermediate stage cell, D = differentiated or dead cell, and M = malignant cell, μ_1 = rate at which I cells are formed from S cells, and μ_2 = rate at which M cells are formed from I cells. The rates α_1, α_2, β_1 and β_2 of cell formation are as indicated. It is assumed that a single M cell can give rise to a tumor.

Table 7.9. Sample Data for Test for Linear Trend

Dose	0	1	2
Number at risk	50	50	50
Number of cases	10	0	20
Percent cases	20	0	40

tions of it, can be expected to play an important role in the understanding of hormesis.

DISCUSSION

The demonstration of hormetic effects is rendered difficult for a number of reasons: The spontaneous rate must be large enough for a difference to be detectable. In contrast with detrimental effects, there is a limited range of doses over which beneficial effects are likely to be found. Publication bias not only hampers publication of low-dose beneficial effects, but discourages research in the area by not providing sufficient motivation for scientific investigators. Some scientists actually believe that hormetic effects are contrary to reason. All these factors contribute to lessen the chances of detecting hormetic effects through synthesis of the scientific literature.

The extra statistical power obtained from mathematical modeling is not available for hormetic studies when appropriate models are not available. Even a simple statistical device such as a test for linear trend does not work well for U-shaped data. For example, with a control group and two treated groups with dose levels of 1 and 2 units, and with 50 animals per group, a test for linear trend on the data in Table 7.9 results in a statistically significant positive trend ($P < 0.02$).

The range of doses exhibiting hormetic effects will generally be too small and too ill-defined to accommodate a test for negative linear trend, since virtually every substance becomes toxic as the dose increases. Yet data such as the above strongly suggest a hormetic effect may exist.

Nevertheless, there are a wide variety of tools available for studying nonmonotonic dose-response curves. As work in this area progresses, more and more substances will be found which exhibit hormetic behavior with one type of health outcome, and perhaps detrimental behavior with another. The situation will become more complex as our knowledge of synergistic and antagonistic relationships between hormetic and nonhormetic substances increases.

REFERENCES

1. Higginson, J. "Editorial: Publication of 'Negative' Epidemiological Studies," *J. Chronic Diseases* 40:371–372 (1987).

2. Furst, A. "Hormetic Effects in Pharmacology: Pharmacological Inversions as Prototypes for Hormesis," *Health Phys.* 52:527–530 (1987).

3. Calabrese, E. J., M. E. McCarthy, and E. Kenyon. "The Occurrence of Chemically Induced Hormesis," *Health Phys.* 52:531–541 (1987).

4. Fears, T. R., R. E. Tarone, and K. C. Chu. "False-Positive and False-Negative Rates for Carcinogenicity Screens," *Cancer Res.* 37:1941–1945 (1977).

5. Downs, T. D, M. M. Crane, and K. W. Kim. "Mortality Among Workers at a Butadiene Facility," *Am. J. Ind. Med.* 12:311–329 (1987).

6. Monson, R. R. "Analysis of Relative Survival and Proportionate Mortality," *Comp. Biomed Res.* 7:325–332 (1974).

7. Monson, R. R. "Observations on the Healthy Worker Effect," *J. Occup. Med.* 28:425–433 (1986).

8. Howe, G. R., A. M. Chiarelli, and J. P. Lindsay. "Components and Modifiers of the Healthy Worker Effect: Evidence from Three Occupational Cohorts and Implications for Industrial Compensation," *Am. J. Epidemiol.* 128:1364–1375 (1988).

9. Bailar, J. C., and F. Ederer. "Significance Factors for the Ratio of a Poisson Variable to Its Expectation," *Biometrics* 20:639–643 (1964).

10. Beaumont, J. J., and N. E. Breslow. "Power Considerations in Epidemiologic Studies of Vinyl Chloride Workers," *Am. J. Epidemiol.* 114:725–734 (1981).

11. Fleiss, J. L., and A. J. Gross. "Meta-Analysis in Epidemiology, with Special Reference to Studies of the Association Between Exposure to Environmental Tobacco Smoke and Lung Cancer: A Critique," *J. Clin. Epidemiol.* 44:127–139 (1991).

12. Mann, C. "Research News: Meta-Analysis in the Breech," *Science* 249:476–480 (1990).

13. Spitzer, W. O. "Editorial: Meta-Meta-Analysis: Unanswered Questions About Aggregating Data," *J. Clin. Epidemiol.* 44:103–107 (1991).

14. Feinstein, A. R. "Scientific Standards in Epidemiologic Studies of the Menace of Everyday Life," *Science* 247:1257–1263 (1988).

15. Downs, T. D., and R. F. Frankowski. "Influence of Repair Processes on Dose-Response Models," *Drug Metab. Rev.* 13:839–852 (1982).

16. Anderson, E. L., and the Carcinogen Assessment Group of the U.S. Environmental Protection Agency. "Quantitative Approaches in Use to Assess Cancer Risk," *Risk Analysis* 3:277–295 (1983).

17. Downs, T. D. "Assessment of Various Dose-Response Models in the Determination of Risk," in *New Approaches in Toxicity Testing and Their Application in Human Risk Assessment,* A. P. Li, Ed. (New York: Raven Press, 1985), pp. 227–233.

18. Moolgavkar, S. H., and D. J. Venzon. "Two-Event Models for Carcinogenesis: Incidence Curves for Childhood and Adult Tumors," *Math. Biosciences* 47:55–77 (1979).

19. Moolgavkar, S. H., and A. G. Knudson. "Mutation and Cancer: A Model for Human Carcinogenesis," *JNCI* 66:1037–1052 (1981).

20. Moolgavkar, S. H. "Carcinogenesis Modeling: From Molecular Biology to Epidemiology," *Ann. Rev. Public Health* 7:151–169 (1986).

21. Thorslund, T. W., C. C. Brown, and G. Charnley. "Biologically Motivated Cancer Risk Models," *Risk Analysis* 7:109–119 (1987).

22. Moolgavkar, S. H., A. Dewanji, and D. J. Venzon. "A Stochastic Two-Stage

Model for Cancer Risk Assessment. I. The Hazard Function and the Probability of Tumor," *Risk Analysis* 8:383–392 (1988).

23. Dewanji, A., D. J. Venzon, and S. H. Moolgavkar. "A Stochastic Two-Stage Model for Cancer Risk Assessment. II. The Number and Size of Premalignant Clones," *Risk Analysis* 9:179–187 (1989).

List of Contributors

William Allaben, National Center for Toxicological Research, Jefferson, Arkansas 72079

Robert E. Anderson, University of New Mexico, School of Medicine, Albuquerque, New Mexico 87131

Harold Boxenbaum, Marion Merrell Dow, Inc., 2110 East Galbraith Road, Cincinnati, Ohio 45215-6300

Ming Chou, National Center for Toxicological Research, Jefferson, Arkansas 72079

Tom Downs, School of Public Health, The University of Texas, Health Science Center, Houston, Texas 77225

Peter Duffy, National Center for Toxicological Research, Jefferson, Arkansas 72079

Emmanuel Farber, Departments of Pathology and of Biochemistry, Medical Sciences Building, University of Toronto, Ontario, Canada M5S 1A8

Ritchie Feuers, National Center for Toxicological Research, Jefferson, Arkansas 72079

Ronald W. Hart, National Center for Toxicological Research, Jefferson, Arkansas 72079

Julian Leakey, National Center for Toxicological Research, Jefferson, Arkansas 72079

Jack Lipman, Hoffmann-LaRoche, Nutley, New Jersey 07110-1190

Beverly Lyn-Cook, National Center for Toxicological Research, Jefferson, Arkansas 72079

Harihara M. Mehendale, Department of Pharmacology and Toxicology, University of Mississippi Medical Center, 2500 North State Street, Jackson, Mississippi 39216-4505

Kenji Nakamura, Life Technology, Gaithersburg, Maryland 20877

Joan Smith-Sonneborn, Zoology and Physiology Department, Box 3166, University of Wyoming, Laramie, Wyoming 82071

Angelo Turturro, National Center for Toxicological Research, Jefferson, Arkansas 72079

Index

Accelerated maturation time, 42
Acclimation, 42
ACDQ. *See* 6-Amino-7-chloro-5,8-dioxoquinoline
Acetamido iminoquinone, 60
Acetaminophen metabolites, 60. *See also* specific types
Adaptation, 113–117, 139–142
Adaptive-physiological view of carcinogenesis, 115–117
Adaptive repair, 139–140
Adenosine diphosphate-ribose, 45
Adenosine triphosphate, 46, 67, 83
Adenylated nucleotides, 45
ADP. *See* Adenosine diphosphate
Adversarial-confrontational view of carcinogenesis, 115
Aflatoxin B, 55
Age-specific mortality rate, 1, 2, 4, 14
Aging, 1, 4
 defined, 2, 10
 DNA repair and, 53–56
 gradualist, uniform view of, 7
 as inevitable, 5
 wear-and-tear theory of, 2
Alarmones, 45
Alcohols, 41, 45, 47, 63. *See also* specific types
Aldehydes, 5
Aliphatic alcohols, 63. *See also* specific types
Alkylating agents, 56. *See also* specific types
Alkyltransferase, 56
Allometric scaling principles, 10–12
6-Amino-7-chloro-5,8-dioxoquinoline (ACDQ), 45
Amosite asbestos, 15, 23–24, 26, 27
Anesthetics, 44. *See also* specific types
Antibiotics, 41. *See also* specific types
Antigens, 101, 102, 103

Antihypertensive agents, 10. *See also* specific types
Antimitosis, 78
Antioxidants, 4, 9. *See also* specific types
Anxiety, 6
Apoptosis/interphase cell death, 98
Appetite abatement, 9
Arsenite, 47
Asbestos, 15, 23–24, 26, 27
ATP. *See* Adenosine triphosphate
Autoimmunity, 106, 109–110
Autoprotection, 62–63, 77
Autoradiography, 72

B-cells, 96, 97, 98, 104, 107
Bioassays, 4
Biochemical studies, 71
Biological disorder, 10
Biomarkers, 42
Bone marrow, 98
Bone marrow-derived cells. *See* B-cells
Bromotrichloromethane, 62, 63, 64

Cadmium, 47
Cadmium chloride, 45
Calcium, 66, 70, 71, 83
Calcium-dependent endonuclease, 106
Caloric restriction, 4, 8, 9, 14, 19, 28, 33
Cancer, 30, 44, 55, 59, 114–117. *See also* specific types
Carbon tetrachloride, 61–62
 autoprotection of, 62–63, 77
 bioactivation of, 61, 62, 63, 66, 68, 69, 70, 71, 78, 80
 chlordecone and, 61, 64–67, 70, 73, 75
  hormesis nonexpression and, 75–76
  two-stage toxicity model and, 79, 82, 83, 84